Status of Recent Geoscience Graduates 2021

Edited by Christopher Keane, Leila Gonzales, and Delaney Robinson

ISBN: 978-0-922152-54-4

Published and printed in the United States of America.

Graphs by Christopher Keane, AGI Workforce Program
Design by Brenna Tobler, AGI Graphic Designer

Christopher Keane
American Geosciences Institute
4220 King Street, Alexandria, VA 22302
www.americangeosciences.org
keane@americangeosciences.org
(703) 379-2480, ext. 219

For more information on the American Geosciences Institute and its publications check us out at **store.americangeosciences.org**.

american geosciences institute
connecting earth, science, and people

Subject reference icons used within this report:

demographics | field of study | curriculum

money | research | social engagement | employment | degree level

Front cover photo: A USGS employee collecting data for the Lake Tahoe Nearshore Nutrient Source Identification Project. Staff collect water quality samples for stable isotope analysis (phosphorus and nitrogen), performing slug tests, and collecting seepage measurements. Photo by Gwen Davies, USGS. Image from the USGS: **https://www.usgs.gov/products/multimedia-gallery**. Icons used in report are from The Noun Project. Identity graphics are copyrighted by AGI.

This report is partially based upon work supported by the National Science Foundation under Grant Numbers EAR-1347209, EAR-1725289, ICER-1740386, ICER-1740844, and ICER-1748780. Any opinions, findings, and conclusions or recommendations expressed in this material are those of the authors and do not necessarily reflect the views of the National Science Foundation. Additional support was provided through the AGI Foundation-funded Center for Geoscience & Society.

Introduction

The 2020–21 National Survey of Recent Geoscience Graduates Exit Survey is the latest in a nearly decade-long effort by the American Geosciences Institute (AGI) to understand the experiences and outcomes of new geoscience degree recipients in the geosciences. This report reflects the results of the survey conducted of graduates in the 2020–21 academic year.

Invitations to complete the survey were sent to all degree-granting geoscience programs in the United States, and additional solicitations were sent to registered student members of several AGI Member Societies. Responses to the survey were voluntary and care has been taken to ensure no individual identifying data is publicly reported.

This year's survey had 441 responses, which represents approximately 7% of geoscience graduates in the United States. This report provides summary views of the results of the survey and reflections on long-term trends for geoscience graduates. Though no specific statistical power or confidence is asserted with these results, continuity of trends over the years and alignment with externally measured trends provides evidence of utility in the insights provided by these results for department chairs, faculty, employers, graduates, and current students.

Of note this year is the addition of questions focused on the development and exposure to specific skills and concepts during a graduate's education. These skills and concepts were identified as part of the Vision and Change in the Geosciences—The Future of Undergraduate Geoscience Education report (**https://www.americangeosciences.org/change**) with input from over 1,000 members of the geoscience community, including employers and faculty.

The format for this year's report has also changed to reflect a data almanac approach. Based on feedback regarding the use of the results of this survey, we are focused on presenting the results as the centerpiece with supportive annotation. Each data topic stands independently. Individual charts will be made available for download from AGI's website. For consistency, when bachelor's or master's degrees are abbreviated as BS or MS, these are inclusive of their BA and MA equivalents.

We would like to extend our appreciation to all the graduates who completed the survey and to the department chairs, faculty, and geoscience society staff who helped disseminate the survey instrument to our graduates.

Contents

1. Demographics

Photo by Craig Whitehead on Unsplash

1.1 Degree Level of Respondents

Degrees Awarded to Respondents, 2020-21

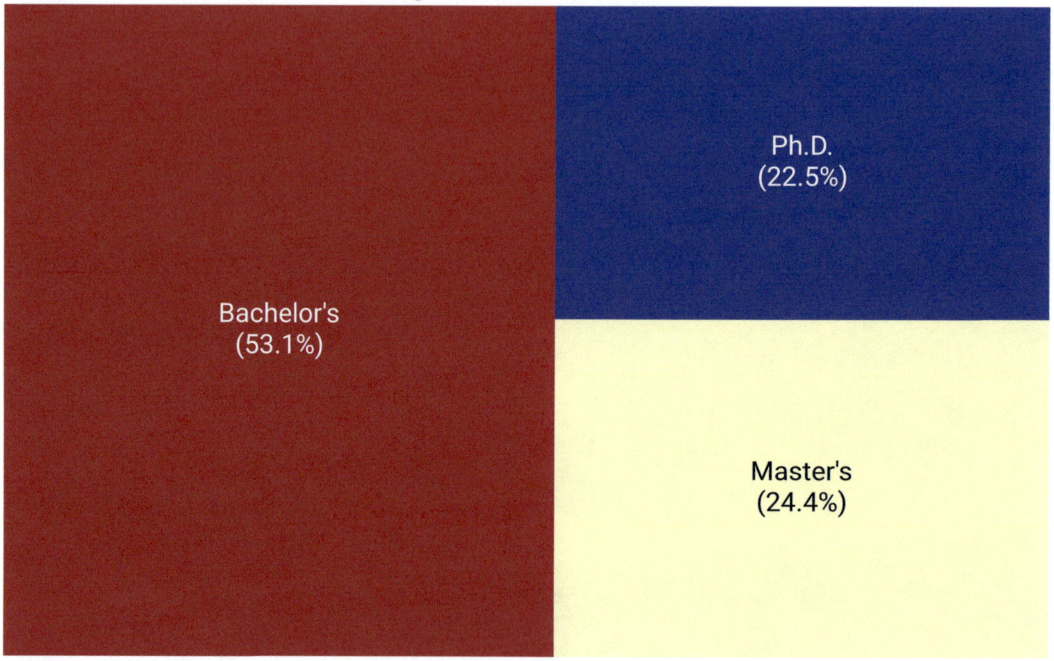

This year, 441 geoscience graduates completed the survey. Just over half of respondents were completing their bachelor's degree, and approximately one quarter each were completing their master's degree or doctorate.

For overall degrees awarded for the geosciences in 2020–21, 66% were bachelor's, 18% were master's and 16% were doctorates, thus the survey responses were underrepresented for bachelor's degrees and overrepresented for graduate degrees.

1.2 Age of Graduates when Degree Awarded

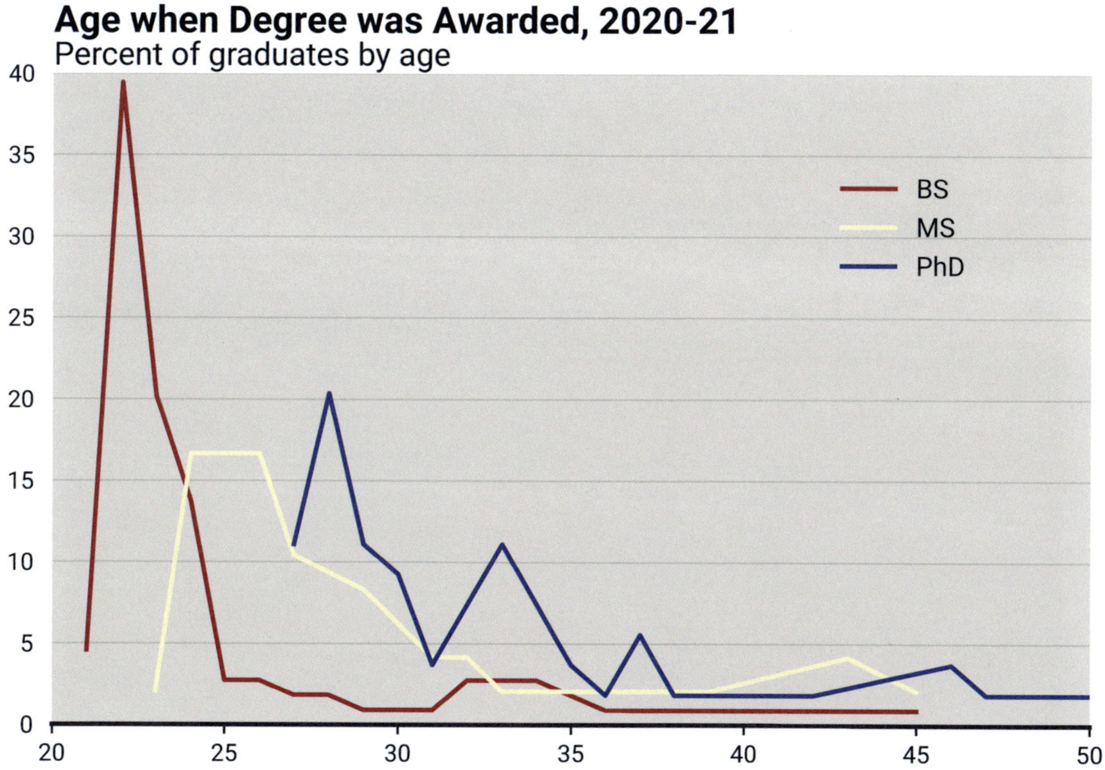

The age distribution of respondents aligns with traditional expectations for each degree level. Most bachelor's graduates had a median age of 23 while master's graduates had a median age of 26.

Doctoral graduates tended to be in their late-20s with a median age of 30 and an average age of 32. This is a minor reduction in the average age of geoscience doctorates which have had a long-term average age at graduation of 33 years old. The driver for geoscience previously having the oldest average age of doctoral recipients was that a large proportion of graduates spent several years in the workforce and then returned to school. Today, more doctoral graduates are continuously in school during their formal education.

1.3 Gender Identity of Respondents

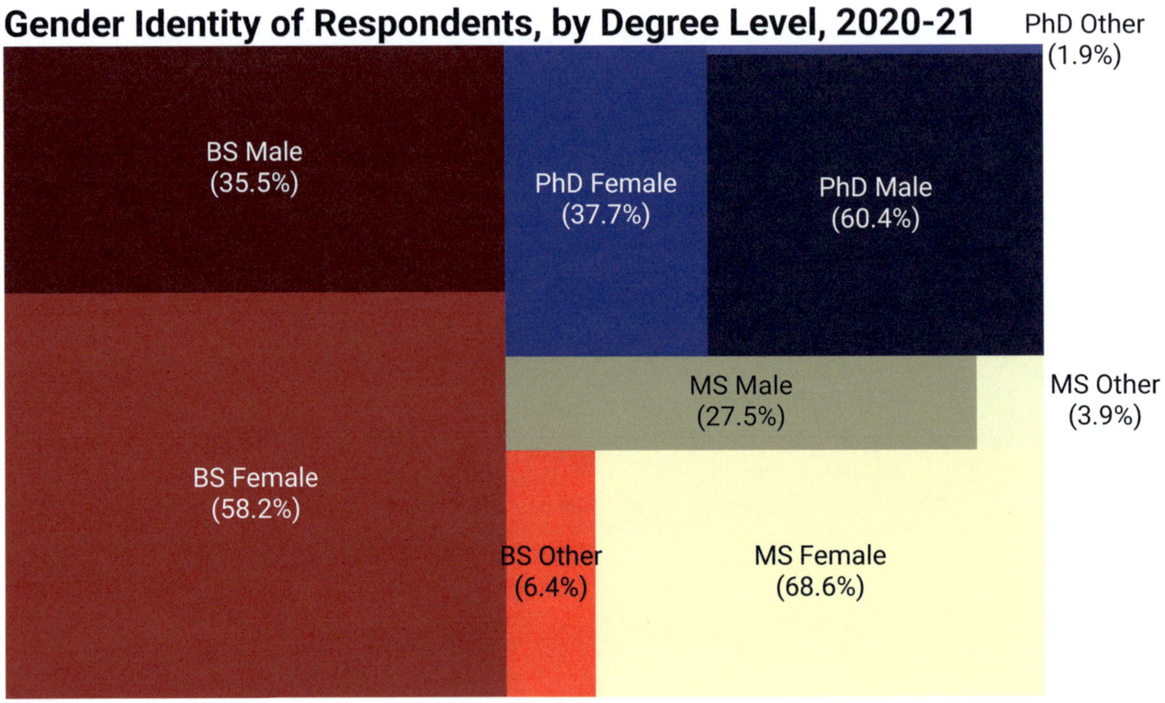

Most survey respondents at both the bachelor's and master's levels were female. The female response rate at the bachelor's level was higher than the most recent survey trends which indicated that just less than half of bachelor's in geoscience were awarded to females.

The response rate of females at the doctoral level is substantially lower than males—which does not align with the functional parity in gender for doctoral degrees awarded.

The response rate for "Other gender" was substantially smaller than the binary gender categories but greater than prior surveys, likely reflecting a cultural evolution in the geosciences.

1.4 Racial Identity of Respondents

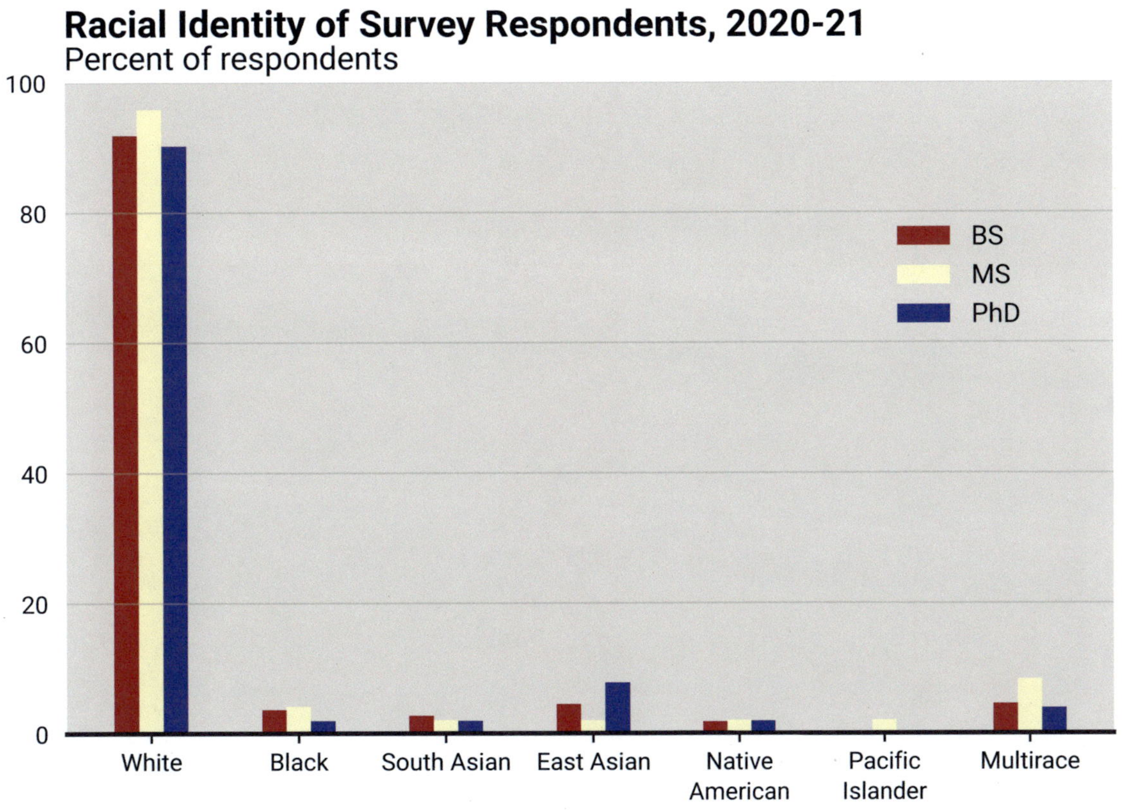

The racial distribution of survey respondents reflects the continuing lack of diversity within the geosciences. Over 90% of respondents classified themselves as White, with nominal responses across other racial identities. Of note is the percentage of respondents reporting as multiracial relative to those reporting a single race. Since being introduced in federal surveys in 2010, multiracial identification has become an important metric in assessing the evolution of diversity in the geosciences. However, the identified component races within the multiracial category are unclear in these results.

1.5 Hispanic Heritage of Respondents

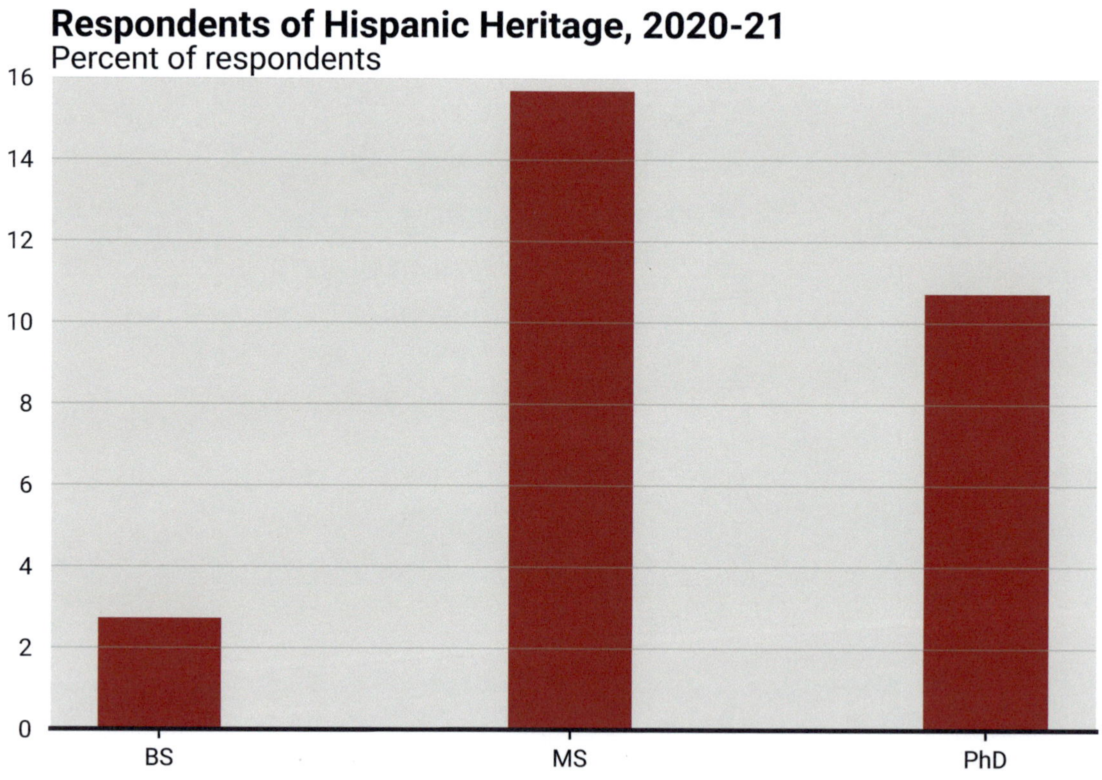

Over the last several years, individuals of Hispanic heritage have represented between 12% and 20% of the respondents in our graduate surveys. This year, even with a record level of responses, Hispanic ethnicity rates collapsed among the bachelor's graduates declining to 3% from 12% in 2017. Since Hispanic ethnicity response rates for master's and doctoral graduates continue to reflect the long-term trends at those degree levels, it is unclear why there was a precipitous drop at the bachelor's level. Three potential causes for this drop are: this is a response artifact for this year's survey; critical support programs for Hispanic geoscience students have ceased; or there is a structural weakness in geoscience programs for supporting the Hispanic community that was aggravated by the pandemic.

1.6 Bachelor's Degree Fields of Study

Field of Degree, Bachelor's, 2020-21

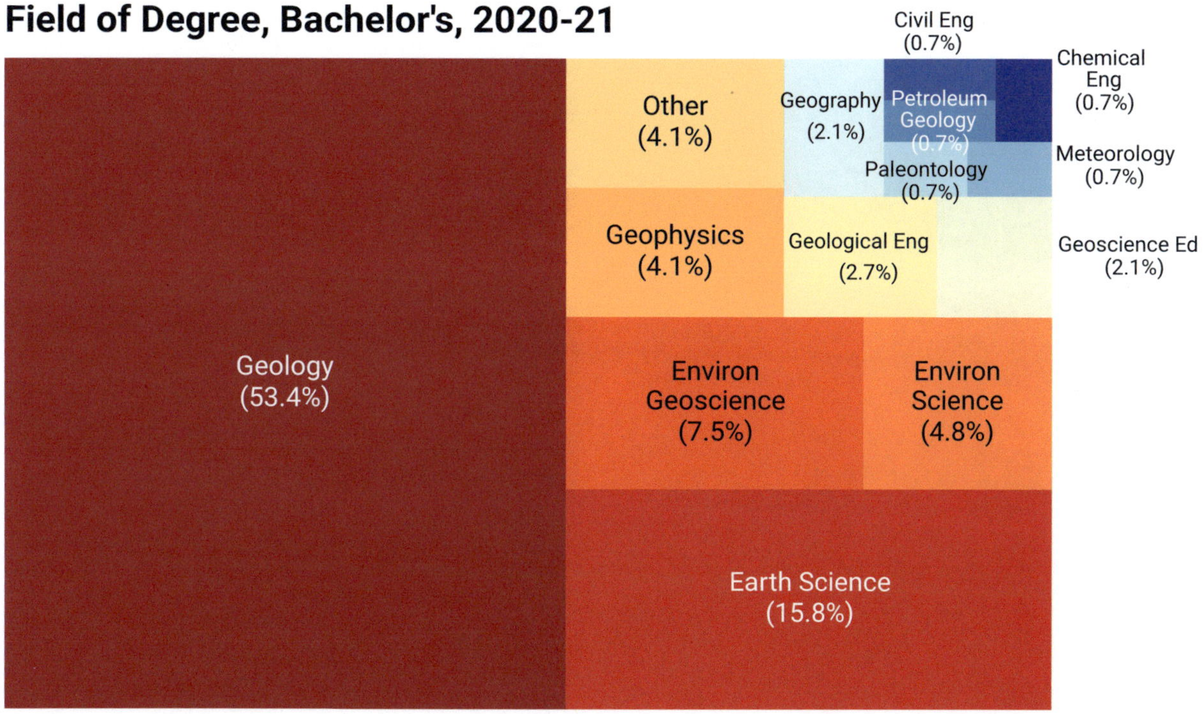

Together, geology and earth science majors comprised almost 70% of the geoscience undergraduate degrees awarded in 2020-2021. The popular topics of environmental geoscience and environmental science constituted slightly more than 10% of degrees, raising questions as to whether the efforts to enroll majors through environmental programs is resulting in additional graduates. More specialized fields of study are smaller fractions of geoscience undergraduate majors, as would be expected at this level since most specialization occurs at the graduate level.

1.7 Master's Degree Fields of Study

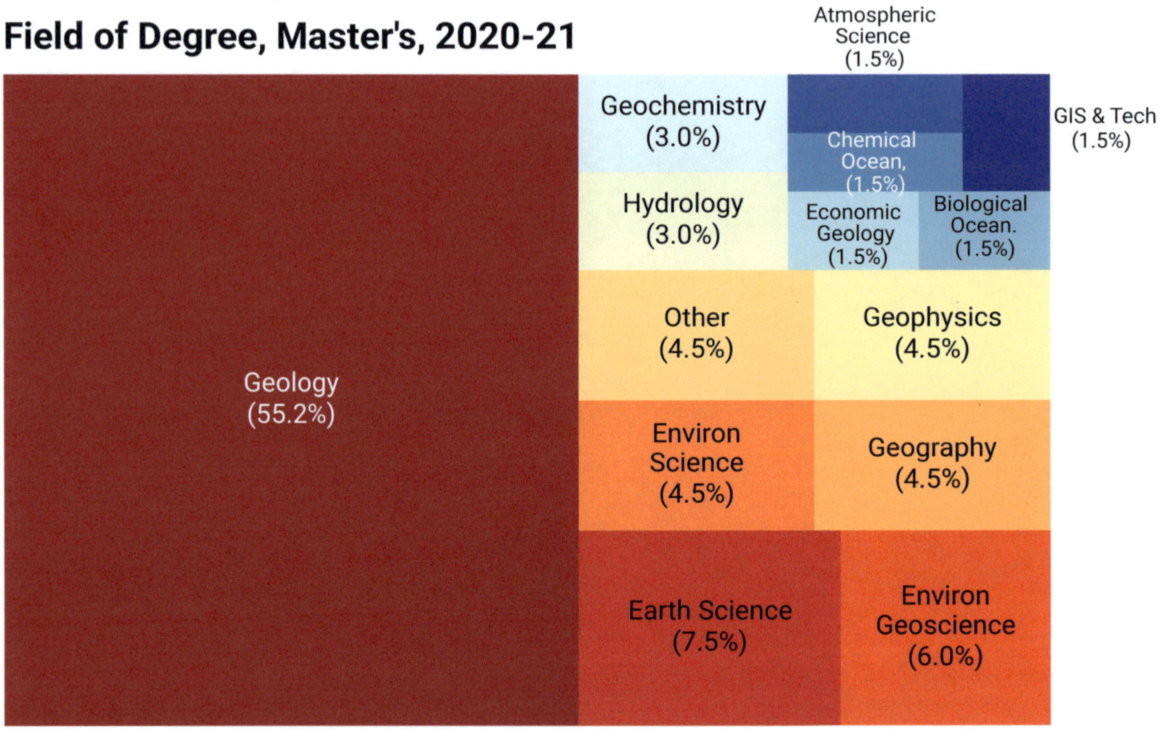

Geology and earth science fields of study represented over half of all master's degrees awarded, with environmental degree fields representing approximately 10% of degrees awarded in 2020-2021. These proportions were similar to bachelor's degrees awarded. As would be expected, there was a slight increase in the diversity of specialty areas at the master's level. Surprisingly, specific areas such as hydrology and GIS did not constitute a larger proportion of master's degrees given that professional services companies and state government agencies have reported seeking these skills in potential new hires, especially in master's graduates.

1.8 Doctoral Fields of Study

Field of Degree, Doctorates, 2020-21

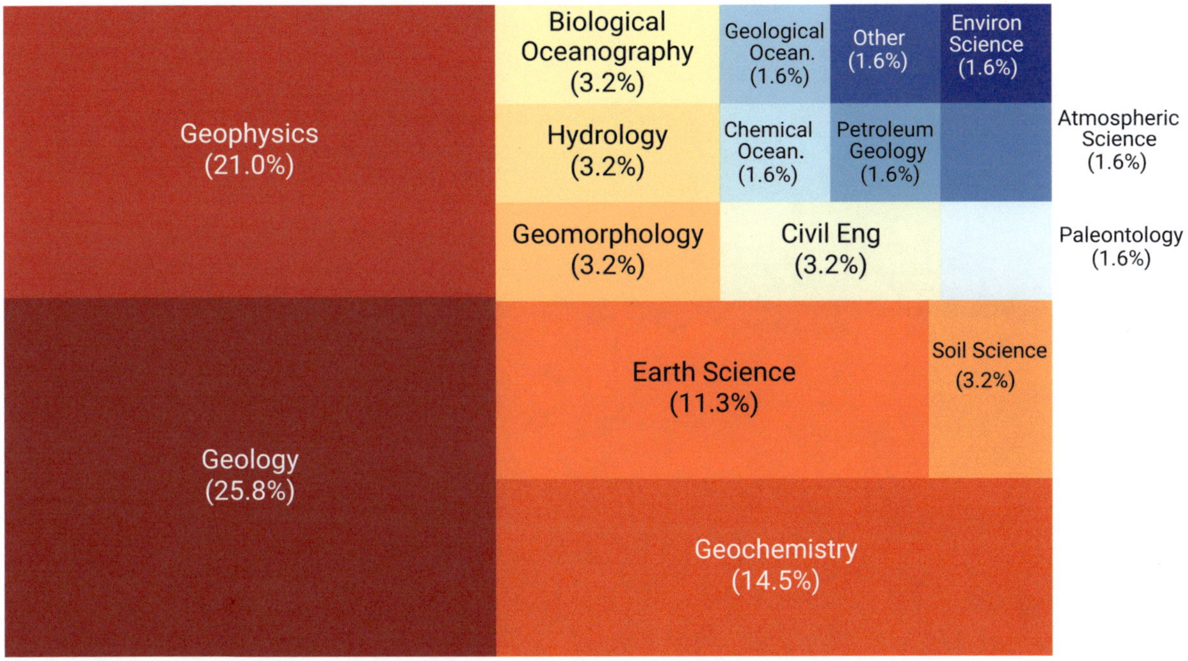

At the doctoral level, geophysics and geochemistry were major fields of study for 2020-2021 graduates while geology comprised a quarter of the doctorates awarded. As expected, with the focused study at the doctoral level, a greater diversity of specialty areas, especially in oceanography and atmospheric science, were evident. Those specialized degree fields built on many of the core geoscience antecedent degrees such as bachelor's and master's degrees in geology.

1.9 Geoscience Enrollment

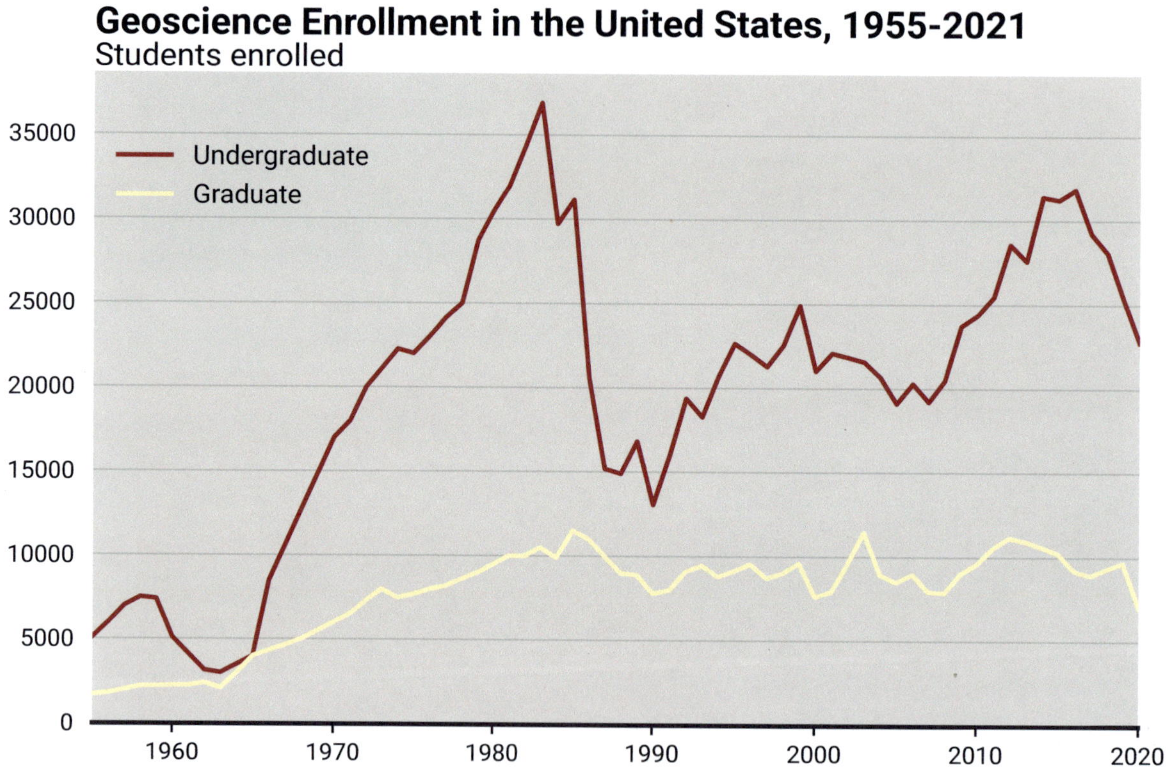

Geoscience undergraduate enrollments continued to decline, a trend starting in 2017 that has been largely driven by a slow-down in oil and gas interest, an increase in the number of major online-only geoscience programs which are difficult to capture in the number of majors, and overall declines in undergraduate enrollments in higher education. A distinct feature of the peak in undergraduate enrollments in the mid-2010s was a large influx of male majors who seemed to be pursuing geoscience degrees to qualify for work in shale-gas operations. This population rapidly exited the workforce when employment in the gas fields contracted.

For the last several decades, graduate enrollments have remained steady and largely at-capacity for graduate programs. The sharp decline this year may reflect COVID-19 pandemic impacts, but the driver will remain unclear until longer trends become apparent.

1.10 **Degrees Awarded**

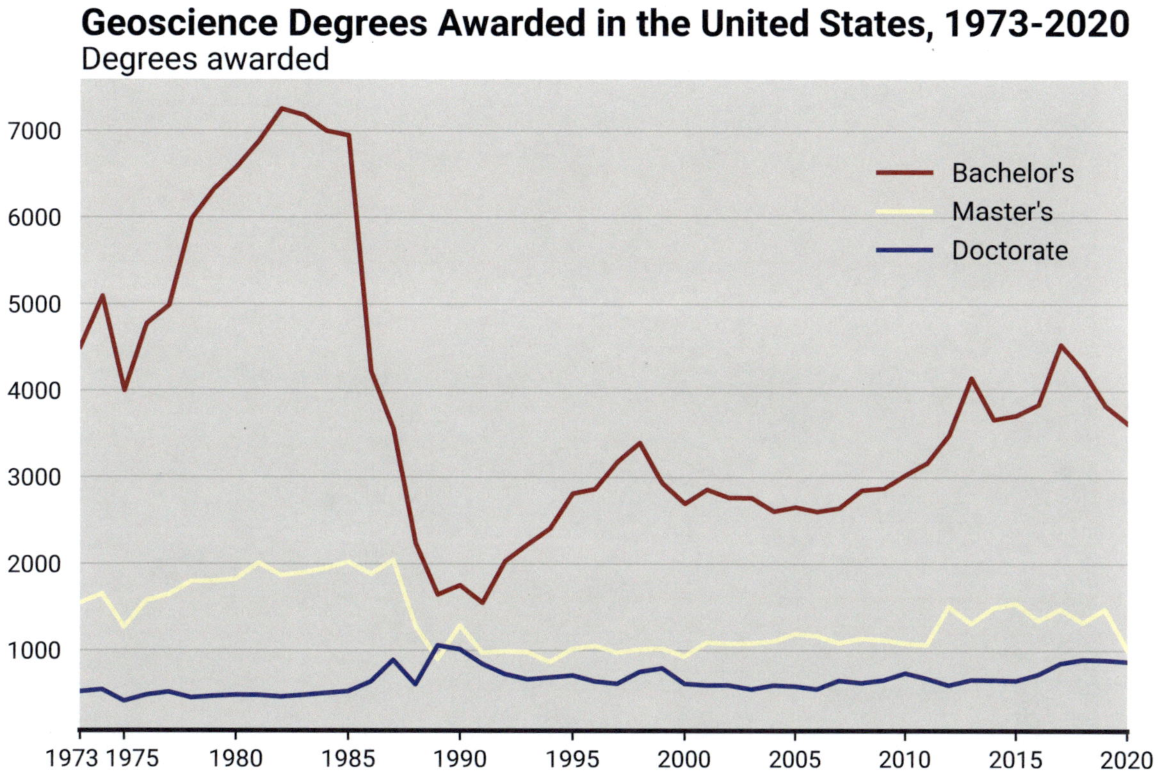

Geoscience bachelor's degrees continued to mirror enrollments with an ongoing decline since 2017. Graduate degrees have remained stable for the last decade, but this year showed a sharp drop in the completion of master's degrees, likely driven by pandemic-related impediments to completion of courses and research.

As seen in this historical perspective, bachelor's degree production is sensitive to labor-market demand for geoscientists, but graduate degrees awarded, especially doctorates, remains generally constant, reflecting the capacity-limited nature of graduate geoscience in the United States.

1.11 **Citizenship of Bachelor's Graduates**

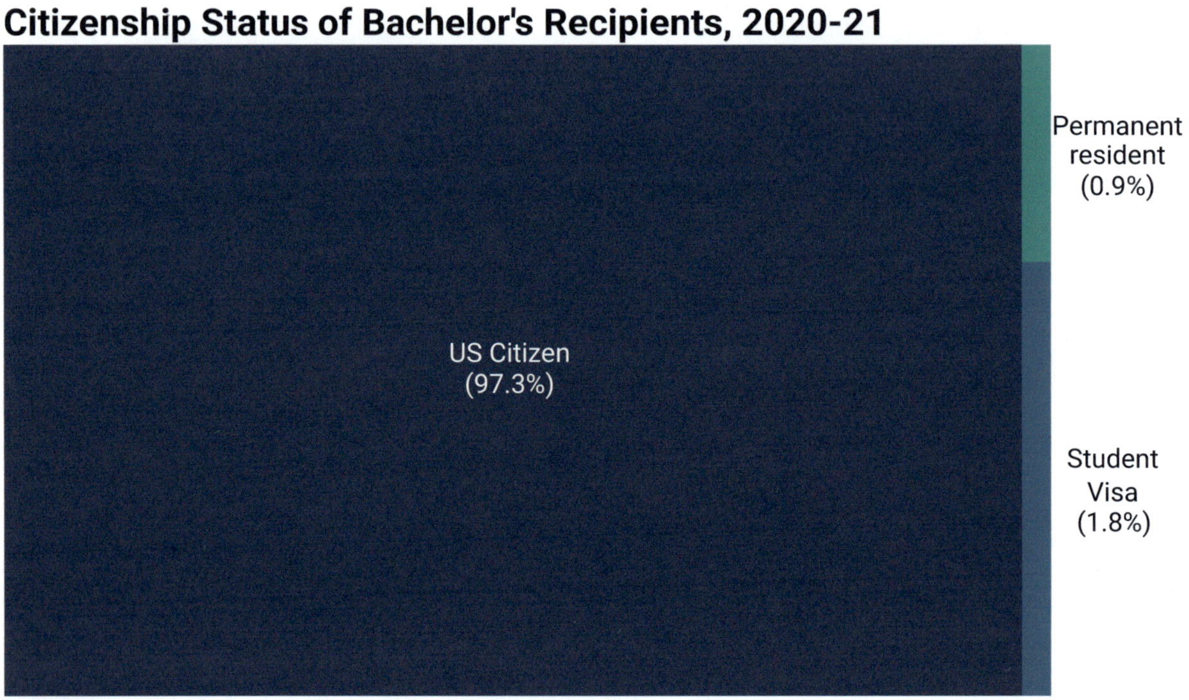

Historically, most U.S. geoscience bachelor's recipients have been citizens or permanent residents. This trend continued into the 2020–21 academic year, with only a very small percentage of graduates who were attending on student visas.

1.12 **Citizenship of Master's Graduates**

Citizenship Status of Master's Recipients, 2020-21

Similar to bachelor's degrees, international students make up a small part of the master's degrees awarded in the geosciences. As an occupation-driven degree, there are few options for non-citizens or green-card holders in the U.S. With the softening of the oil and gas industry, fewer international employees are being educated in the U.S. as well.

1.13 Citizenship of Doctoral Graduates

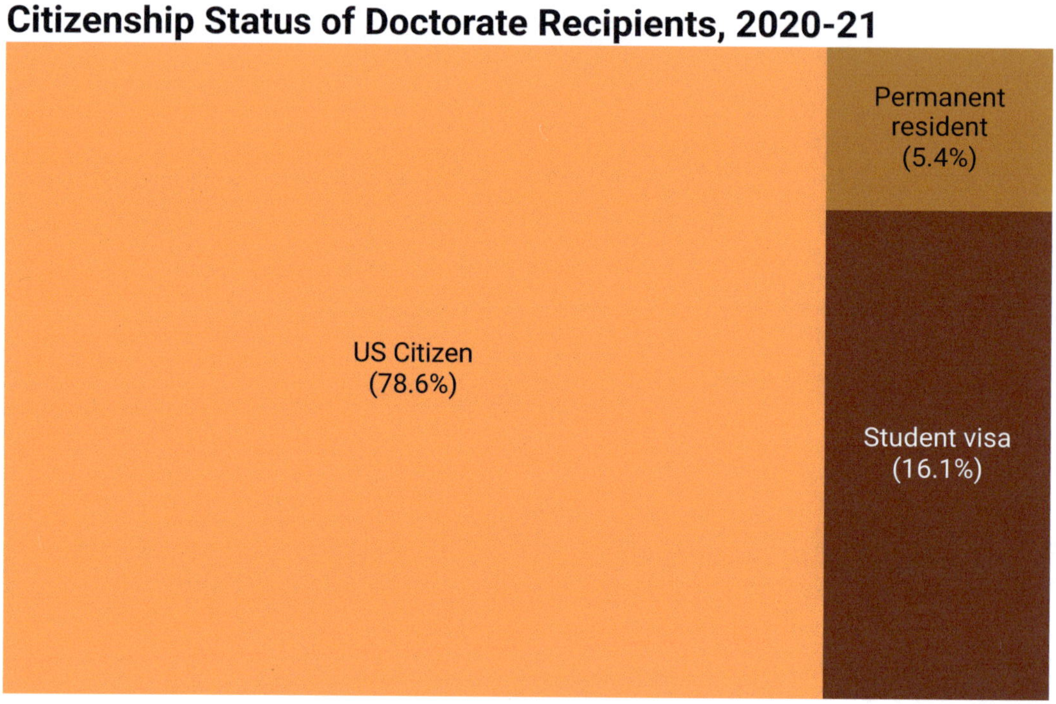

Like most degree fields in the U.S., the doctoral level has the greatest level of participation by non-U.S. citizens. Compared to science and engineering where approximately 34% of doctorates were awarded to non-U.S. citizens, non-U.S. citizen geoscience doctoral students remained a minority in 2020-2021, earning only 16.1% of doctoral degrees.

The drivers for this lower rate of international student engagement within the geosciences are likely complex, including challenging U.S. immigration policies, impacts from the COVID-19 pandemic, weak oil and gas hiring, and that the geosciences have many world-class non-U.S. graduate programs.

1.14 When Graduates Decided to Pursue Geoscience

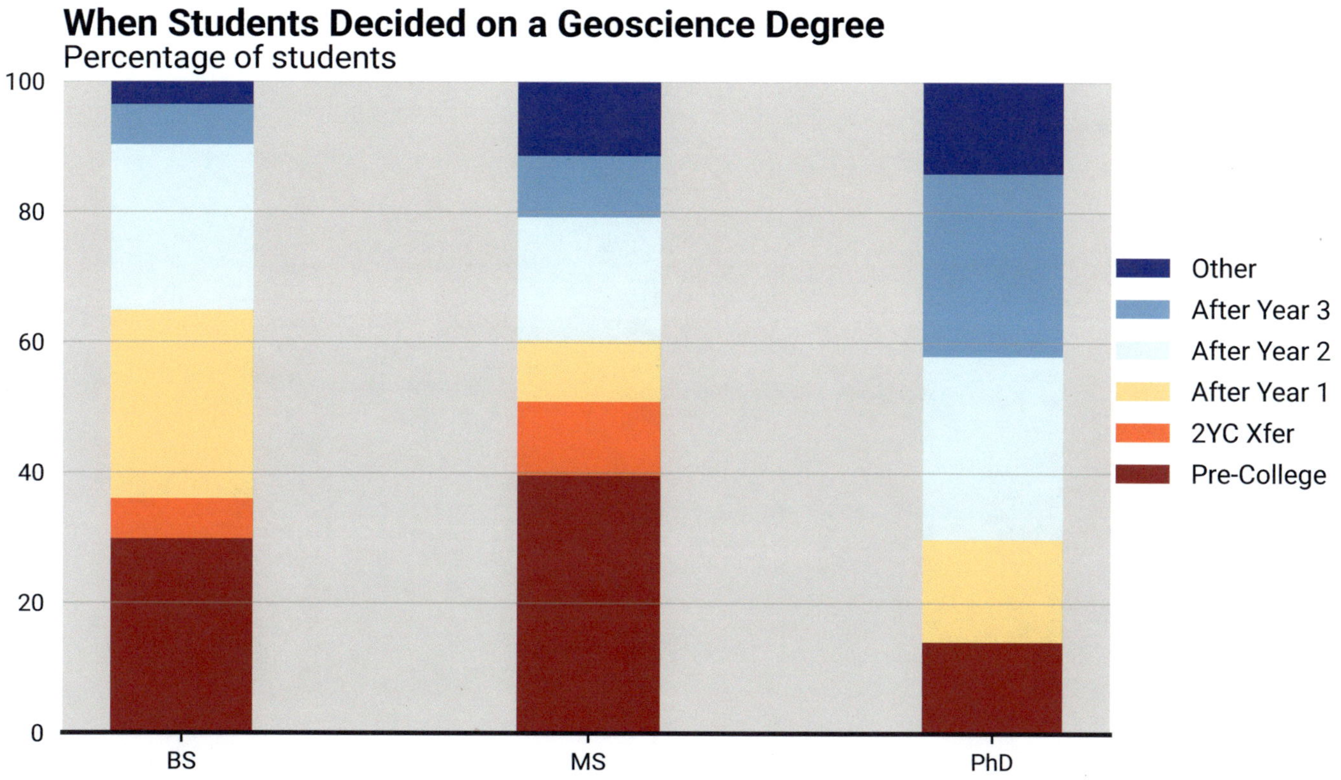

An ongoing discussion regarding student recruitment has centered on when individuals decide to major in the geosciences. Understanding when degree recipients decided to major in the geosciences can identify critical success points. Approximately 60% of bachelor's graduates decided to major in geoscience by the end of their first year in college, similar to the rate of master's graduates. Doctoral graduates tended to make their decision later, reflecting the greater diversity of educational backgrounds which lead into geoscience doctoral programs.

Of note is that 50% of master's graduates decided on geoscience either before entering college or as a transfer student from a 2-year college. This early decision propensity by master's graduates may indicate an affinity to enter the profession with earlier engagement in geoscience.

1.15 **Exposure to Earth Science in High School**

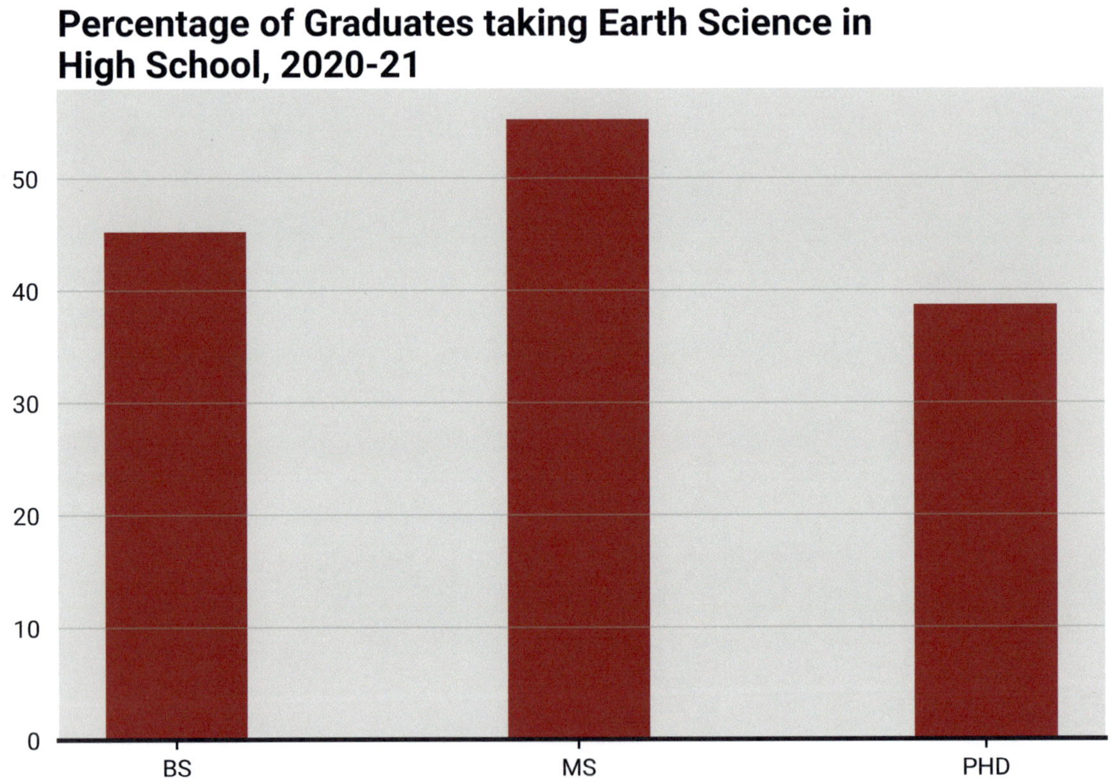

Approximately 50% of graduates took an earth science course in high school. The greatest proportion was at the master's degree level which is also the group with the most individuals deciding to major in geoscience prior to college. This association may support assertions that exposure to geoscience in secondary education has a potential positive influence on individuals eventually becoming practicing geoscientists, or that early disposition to the geosciences remains persistent throughout secondary and higher education.

1.16 Exposure to Geoscience in Community College

Percentage of Graduates with Geoscience Coursework at a Community College, 2020-21

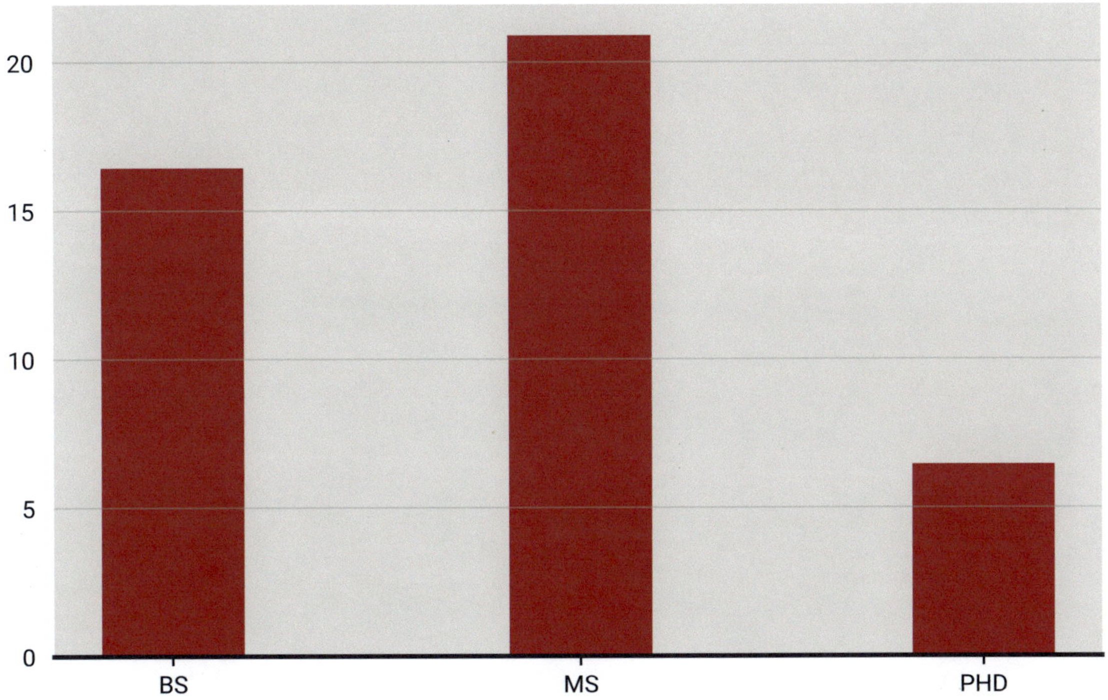

Two-year college is a major part of the educational pathway for new bachelor's graduates as well as for graduate degree recipients. Bachelor's and master's graduates were more likely to have had a geoscience course at a 2-year college than were doctoral graduates. Whereas 16% of bachelor's and 21% of master's graduates took a geoscience course in a 2-year college, only 6% of doctoral recipients did so. The rate for doctoral recipients declined from prior years where approximately 9% of doctoral recipients took geoscience courses at 2-year colleges. A likely driver of this declining trend for doctorate recipients is the diverse disciplinary pathways of many of these graduates.

1.17 Educational Background of Graduates' Parents

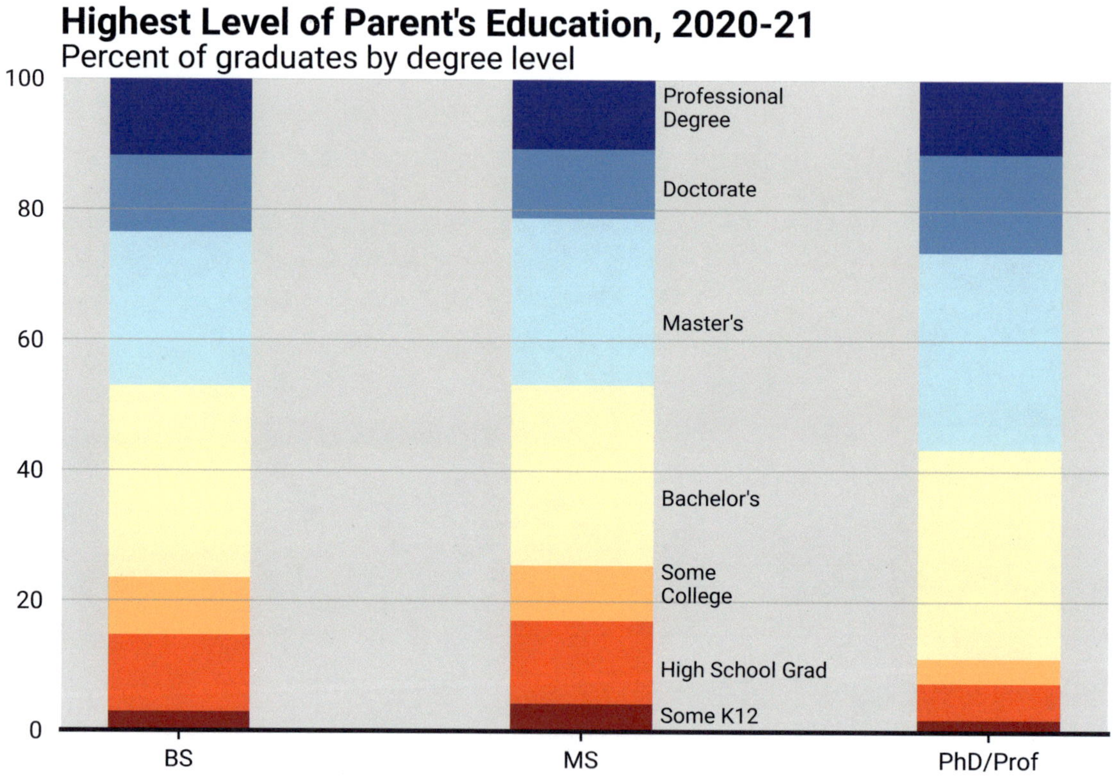

The geosciences have one of the lowest participation rates of first-generation college students. This year however, 22% of bachelor's recipients reported that their parents did not graduate college, which is a substantial increase from only 9% of graduates being first-generation graduates in 2017. With approximately 29% of all bachelor recipients in the U.S. being first-generation graduates, geoscience growth in first-generation student participation is promising.

At the bachelor's and master's level, the educational backgrounds of graduates' parents were similar. At the doctoral level, however, very few graduates were first-generation students, and this is potentially attributable to either the previously low rate of participation by first-generation students in geoscience academic programs or the potential for less familiarity of first-generation students to the culture and experience of doctoral education.

2. Curriculum

Photo by Dom Fou on Unsplash

2.1 Geoscience Concept Exposure for Bachelor's Graduates

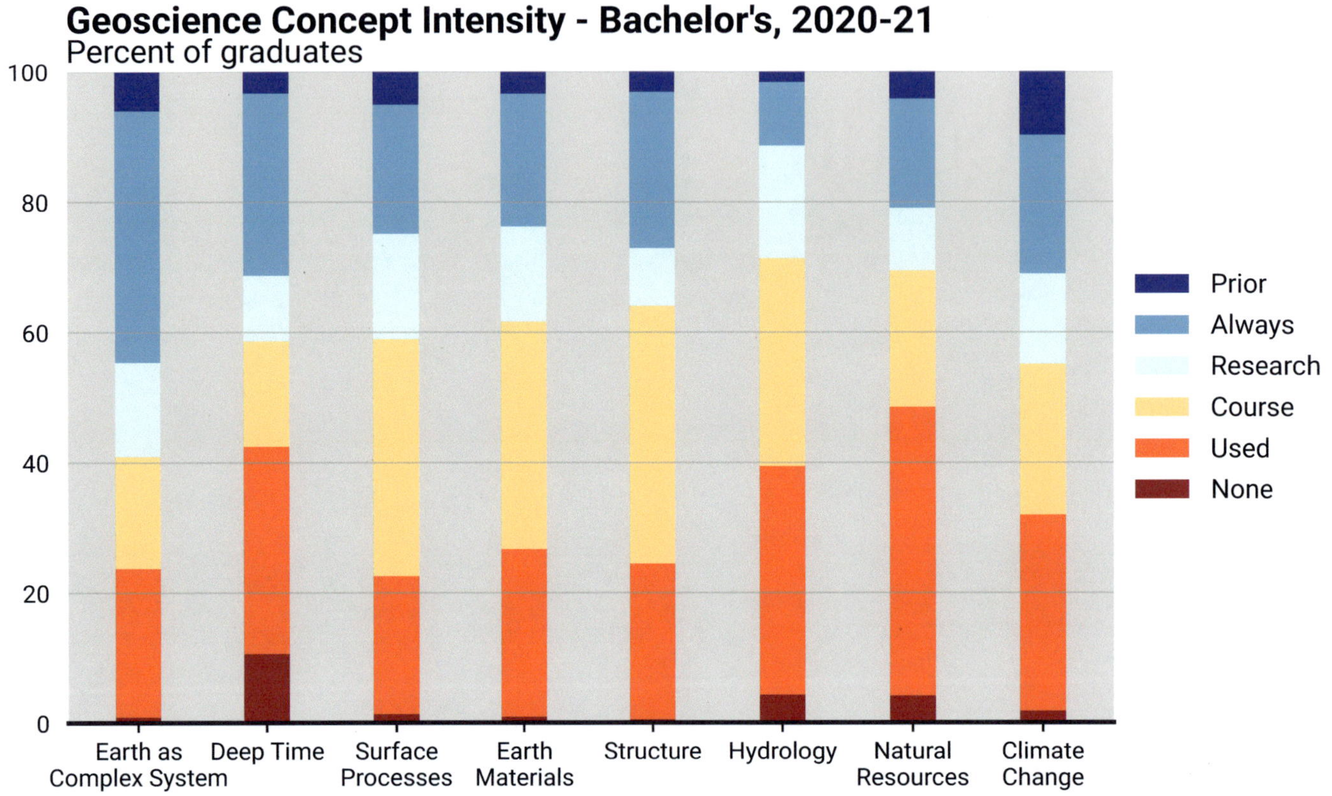

Graduates were asked about their exposure to geoscience concepts that align with the recommendations in the Vision and Change Report. At the bachelor's level, most major geoscience concepts were introduced through a course or used in research or other persistent modes. Certain areas had weaker exposure in that they were not formally introduced to more than 40% of degree recipients, namely deep time, hydrology, and natural resources. Interestingly, the single most intensively introduced topic that bachelor's graduates reported was Earth as a complex system, representing a clear shift in how programs have begun to address the geosciences through an earth systems framework.

2.2 Geoscience Concept Exposure for Master's Graduates

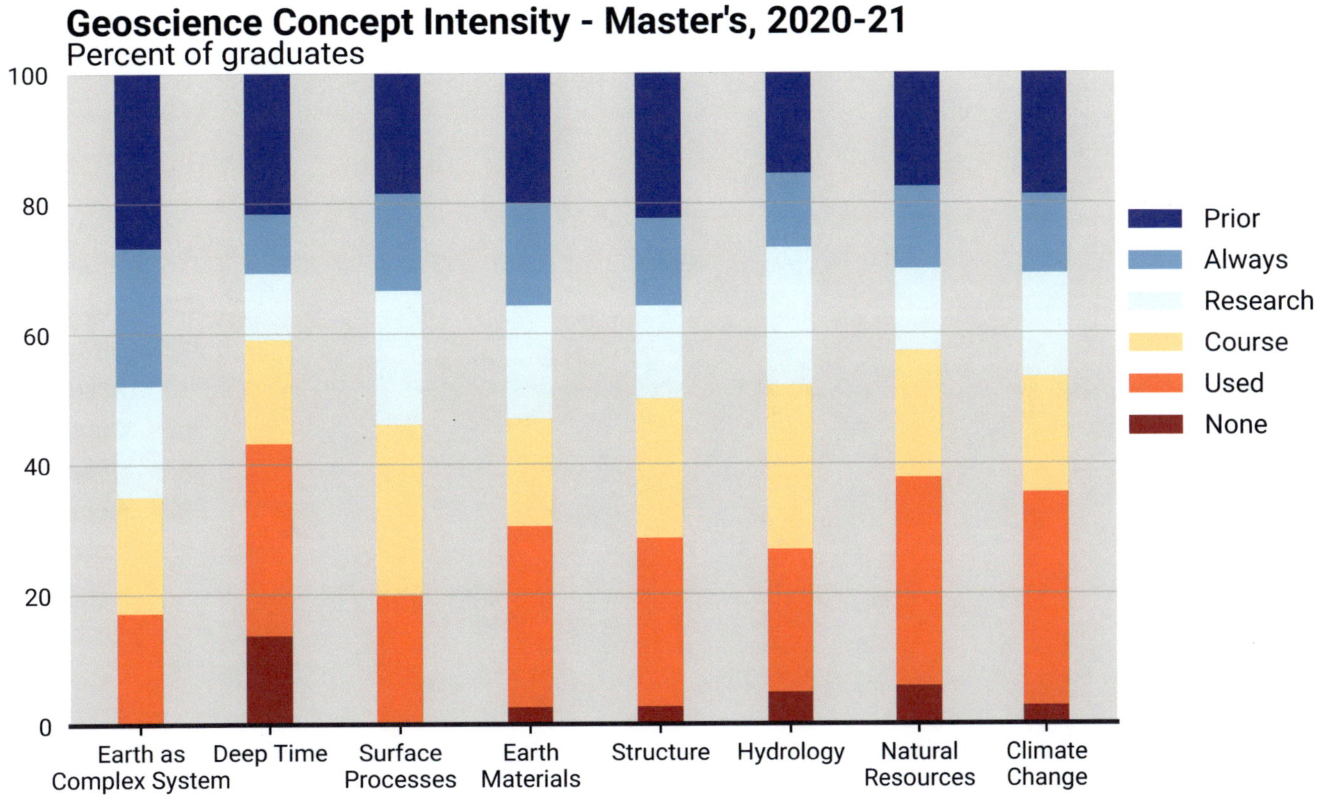

As would be expected, master's graduates showed greater intensity of exposure across all geoscience concepts. Interestingly, the single weakest topic was that of deep time which likely reflects the focus in many master's programs on environmental and near-surface processes that may not be as dependent on engaging in traditional deep-time geologic contexts.

2.3 Geoscience Concept Exposure for Doctoral Graduates

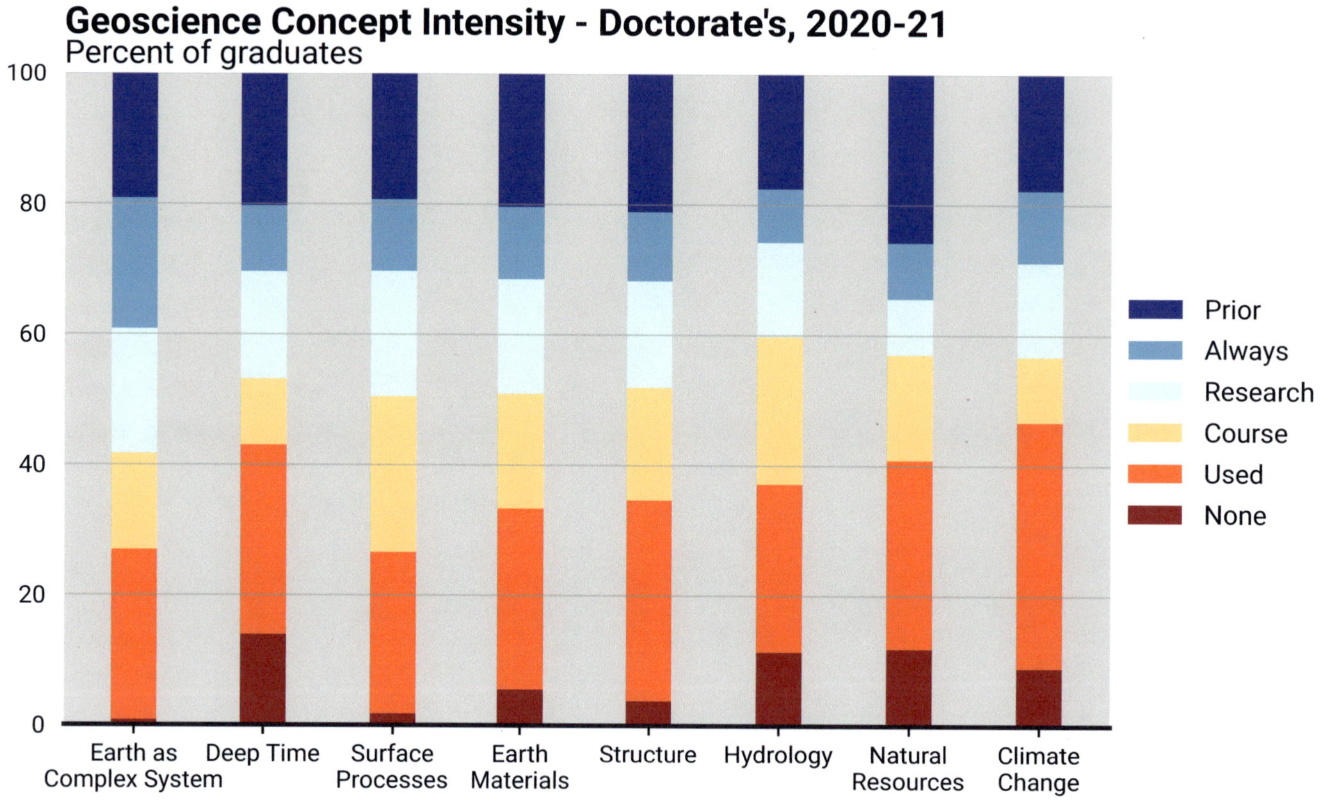

Similar to master's graduates, doctoral graduates had greater intensity of engagement across most geoscience concepts than bachelor's graduates. Of note is that nearly 10% of doctoral graduates reported non-exposure to deep time, hydrology, natural resources, or climate change concepts. These concepts were also only formally addressed with about 60% of graduates. This may reflect two factors: doctoral studies tend to be narrowly defined and therefore these topics may not have been pertinent to graduates' courses of study, and there is a pronounced population of doctoral graduates who came to the geosciences from other disciplines in which some of these foundational topics may not have been introduced.

2.4 Geoscience Skill Levels of Bachelor's Graduates

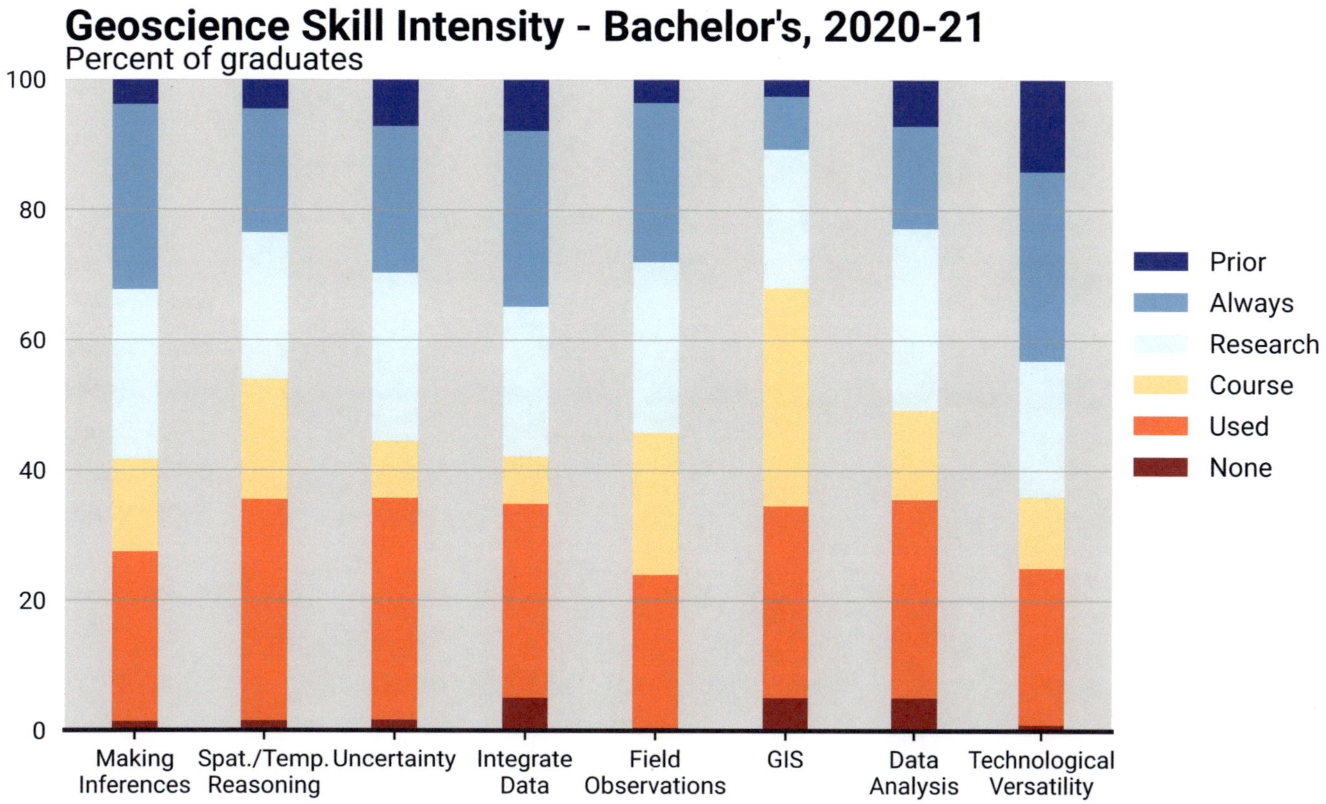

Bachelor's graduates were asked about their level of exposure to a range of geoscience skills identified in the Vision and Change report. Most respondents indicated at least some level of exposure to all of the geoscience skills and well over 50% indicated a fairly intense exposure to all the skills, with only data integration, GIS, and data analysis techniques reflecting any level of non-exposure.

2.5 Geoscience Skill Levels of Master's Graduates

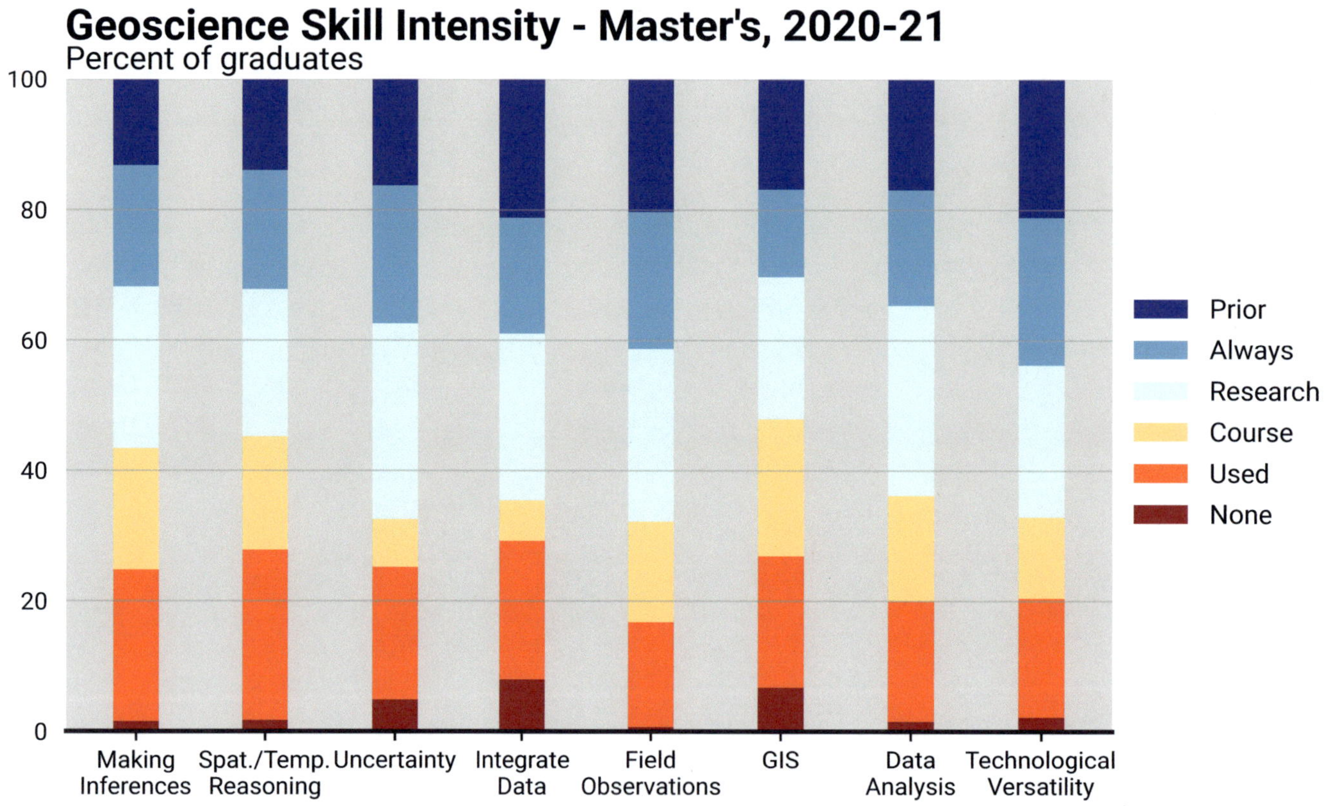

At the master's level, most graduates reported formal exposure to the range of geoscience skills, and in general at greater intensity levels than did bachelor's graduates. The topics of data integration and GIS were the only skills with any notable level of non-exposure; however, nearly 20% of master's graduates indicated prior experience with data integration, field observations, GIS, data analysis and technological versatility.

2.6 Geoscience Skill Levels of Doctoral Graduates

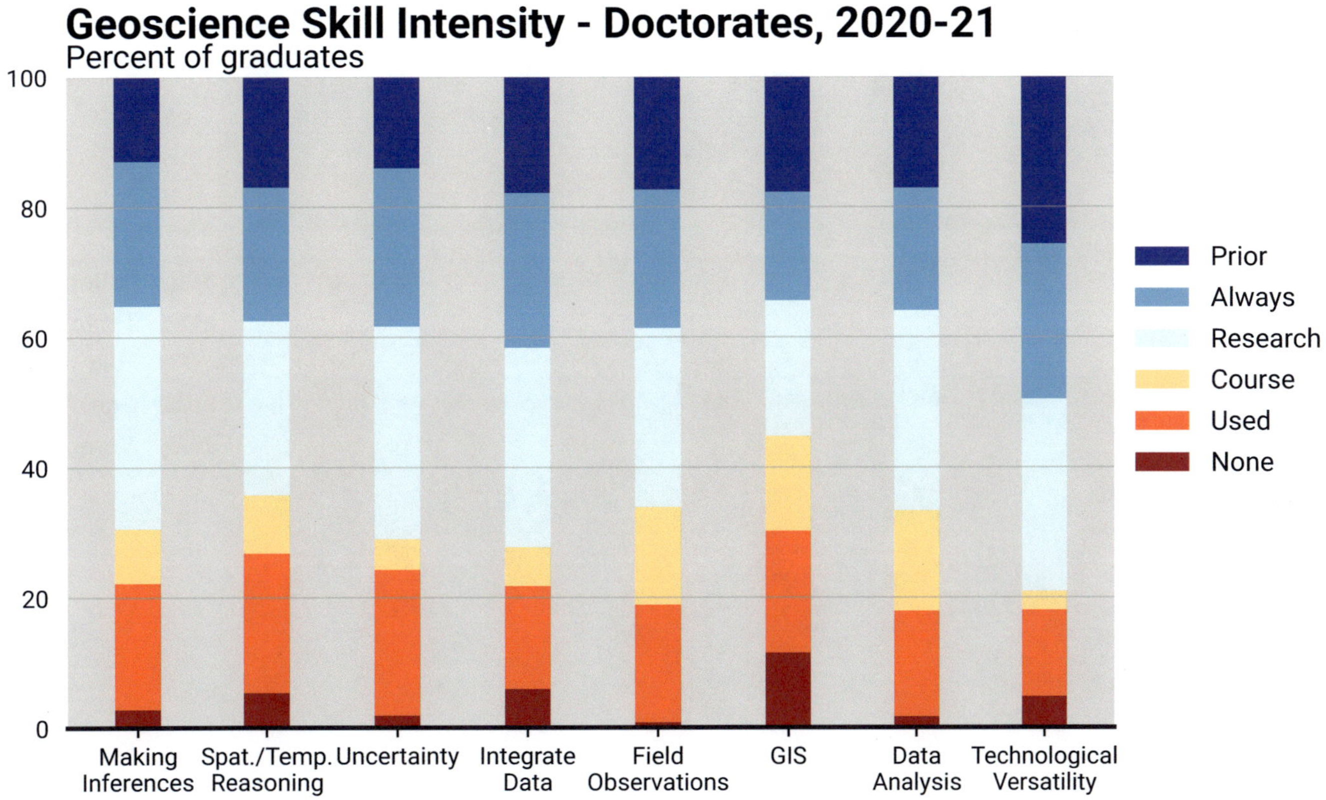

As expected at the doctoral level, most geoscience skills showed either prior mastery, consistent use, or active use within research. Geoscience skill exposure at the doctoral level did not vary substantially from master's graduates.

However, certain areas such as GIS showed a notable lack of exposure potentially representing a legacy difference given curricular changes since many of these doctoral students were undergraduates.

2.7 Field Courses Taken

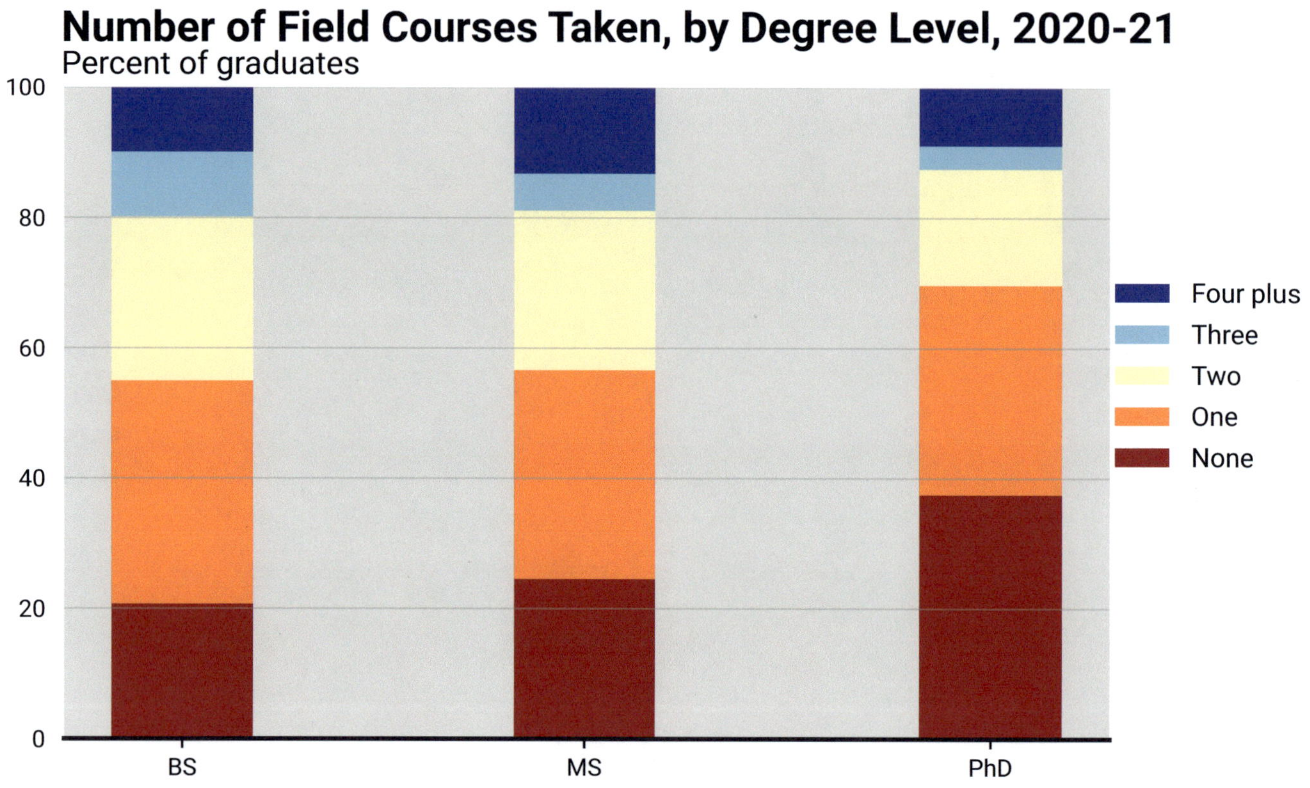

Traditionally, dedicated field courses have been a cornerstone of many geoscience degree programs. At the bachelor's and master's level approximately 80% of students had at least one formal field course, and approximately 60% of doctoral students had a formal field course as well. These results are consistent with the long-term trend over the last 10 years, with lesser field experience at the doctoral level representative of both the diverse background of geoscience doctoral students and that the specific research topics may not require dedicated field experiences.

2.8 Field Courses Taken by Gender

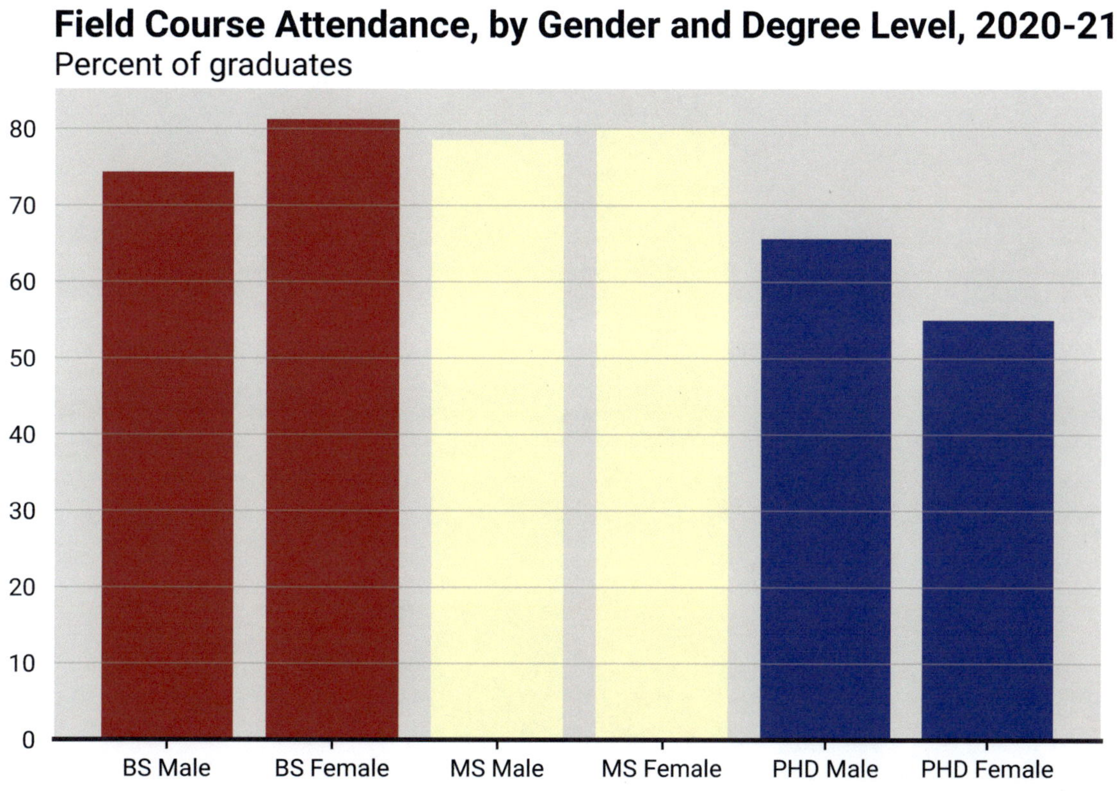

With the continuing growth in female participation in geoscience degree programs, are there any gender-dependent differences in taking formal field courses? For bachelor's recipients, a greater percentage of females had at least one formal field course than did males, and at the master's level, there was minimal difference. The more pronounced difference at the doctoral level may reflect temporal artefacts or potential antecedent degree differences by gender, which was not examined in this study.

2.9 Field Experiences in Courses

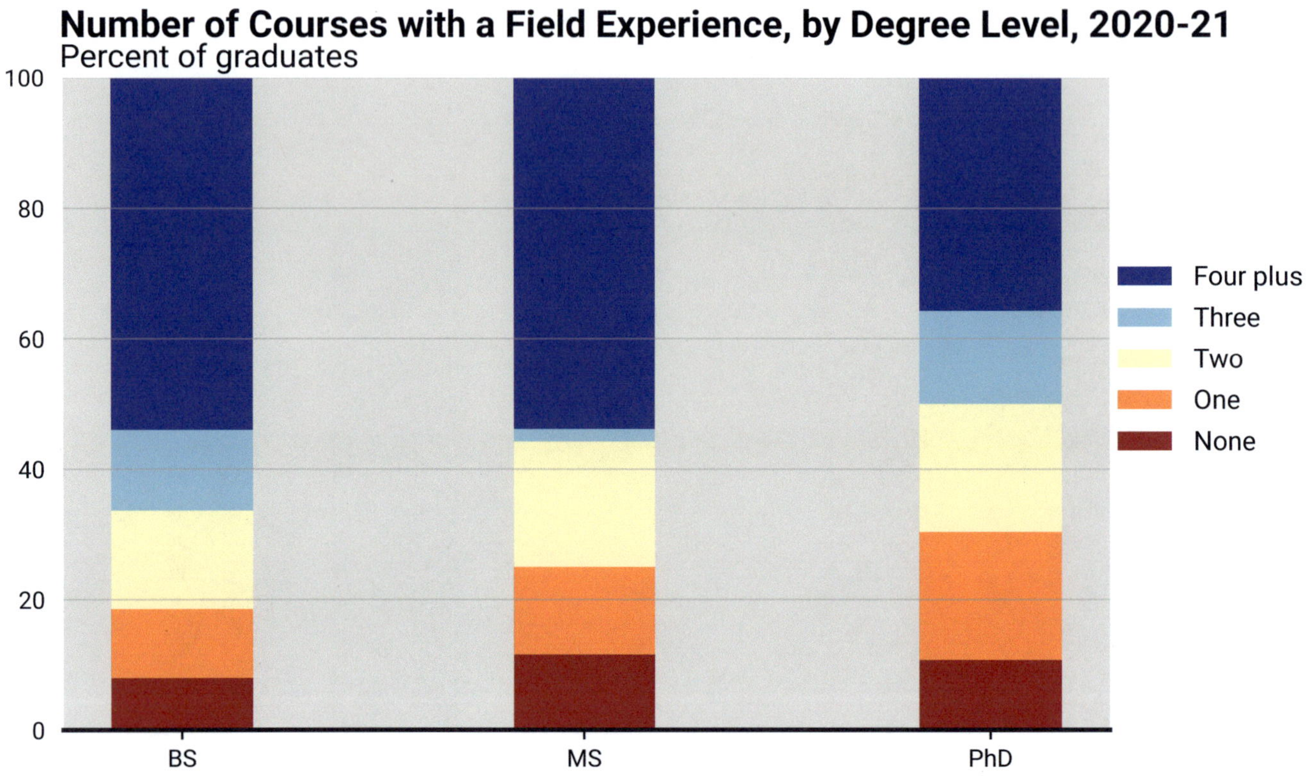

Across all degree levels, at least 90% of graduates reported having at least one course with an integrated field experience. About half of all bachelor's and master's graduates, and over a third of doctoral graduates, indicated at least four or more courses that had a field experience. This trend aligns with the consensus of graduates in this and AGI's COVID impacts study that their field skills and exposure was excellent across all geoscience programs, as reflected by their employment experiences after graduating. Even with the pandemic and the introduction of concepts such as virtual field trips, the core importance and quality of field instruction remains prevalent across all geoscience programs.

2.10 Science Skills of Bachelor's Graduates

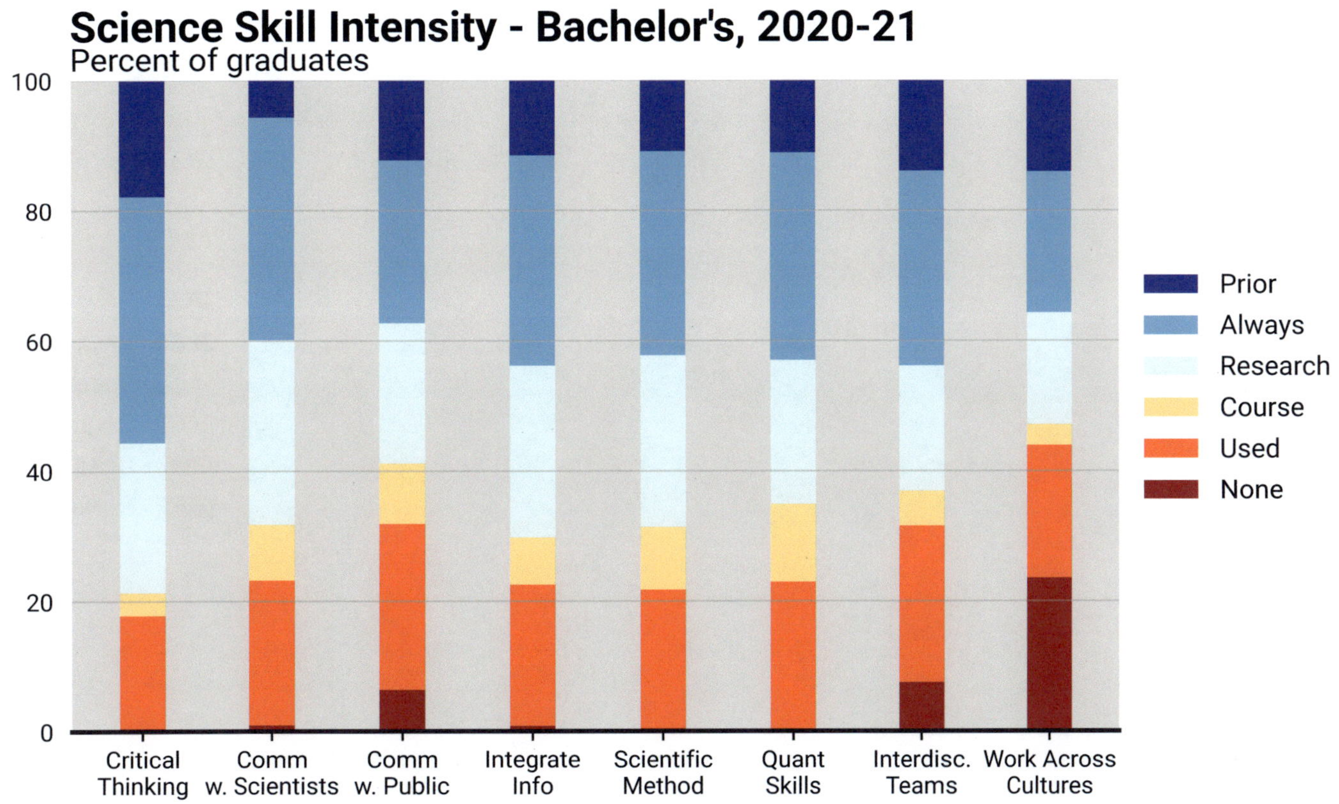

In general, about 75% of bachelor's graduates reported having formal exposure to a wide range of core science skills as defined in the Vision and Change report. The only areas with notable weakness among these graduates were working across cultures, working in interdisciplinary teams, and communicating with the public. The lower level of intensity for working across cultures, and to a lesser degree communicating with the public are not surprising given the typical structure of geoscience degree programs. However, the reported lack of exposure to interdisciplinary teamwork is curious as 70% of graduates indicated some level of formal engagement. We are unable to ascertain whether the graduates were lacking any exposure to working in interdisciplinary teams because of issues such as pandemic-driven remote learning, because it was absent from their curriculum, or because they did not recognize the skill within their educational activities.

2.11 **Science Skills of Master's Graduates**

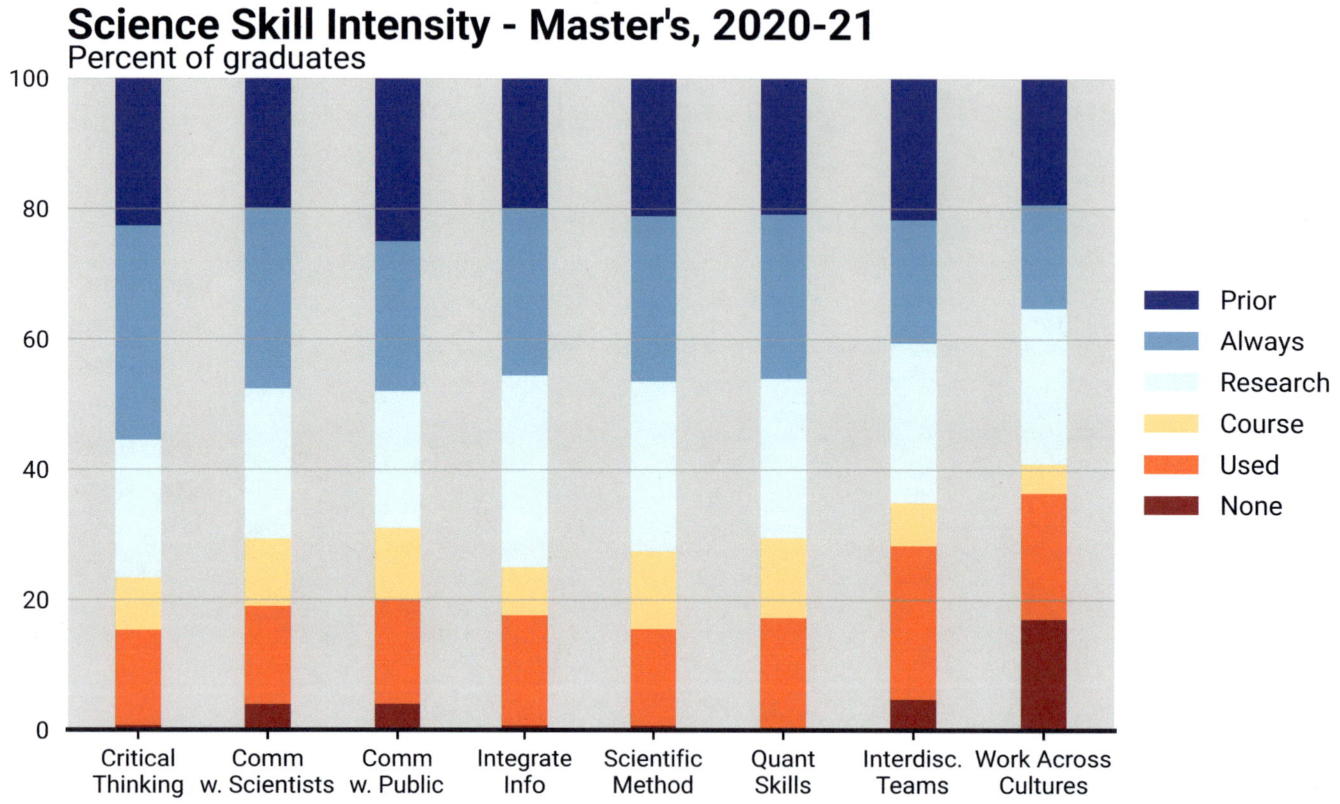

Nearly 50% of master's graduates reported either prior experience or continuous use of many of the science skills identified in the Vision and Change report. This trend is supported by employer assertions that geoscience graduate degree recipients tend to be excellent scientists. The exposure to science skills for master's graduates is similar to bachelor's graduates with notable weakness in working across cultures and in interdisciplinary teams. These deficiencies could be driven by programs that do not rely on cohorts, collaborative work, and team projects, but are strictly structured around independent study and research.

2.12 **Science Skills of Doctoral Graduates**

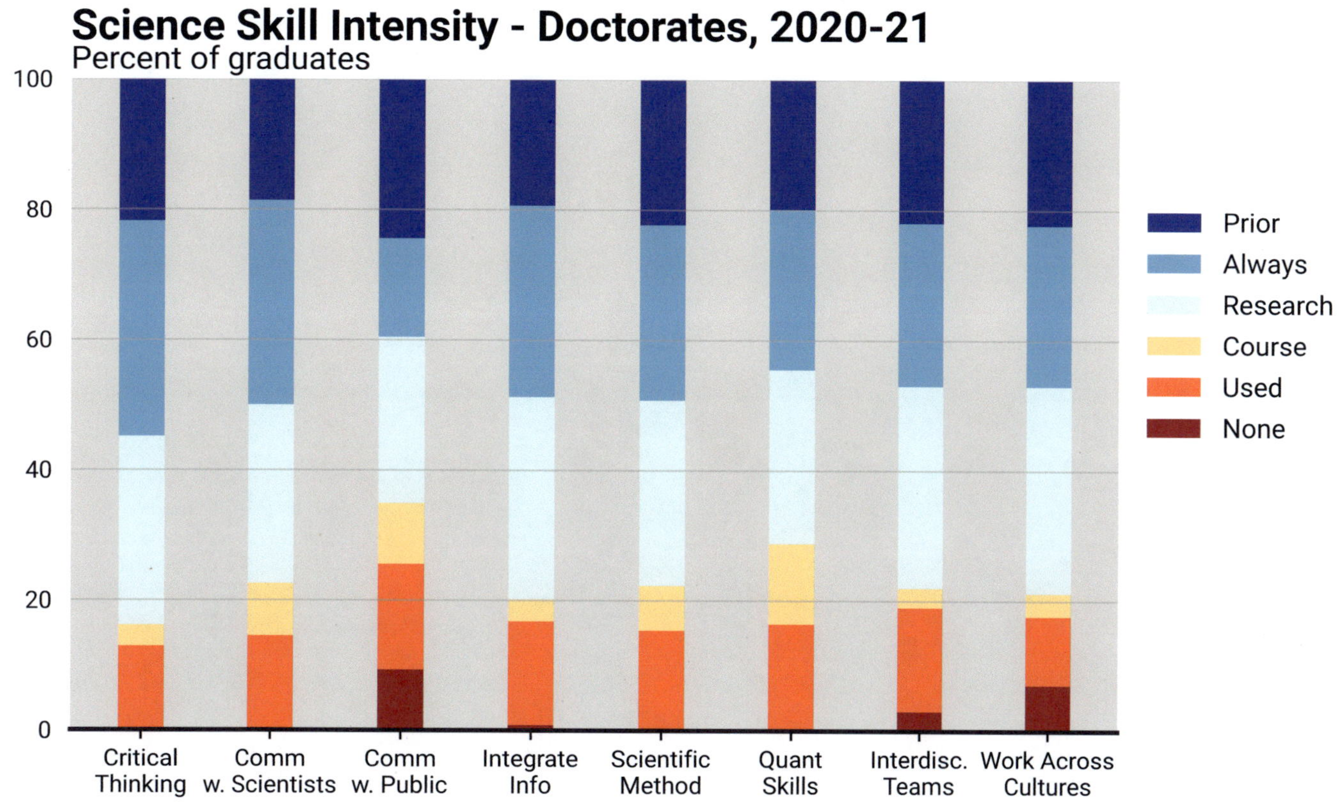

Doctoral graduates reported extensive exposure and utilization of core science skills. The only areas of weakness were in communicating with the public and working across cultures. Those graduates reporting no exposure to these skills may have been working on research not directly aligned with these skills and may not have been introduced to these skills during their course of study.

2.13 Complementary Science Courses by Degree Level

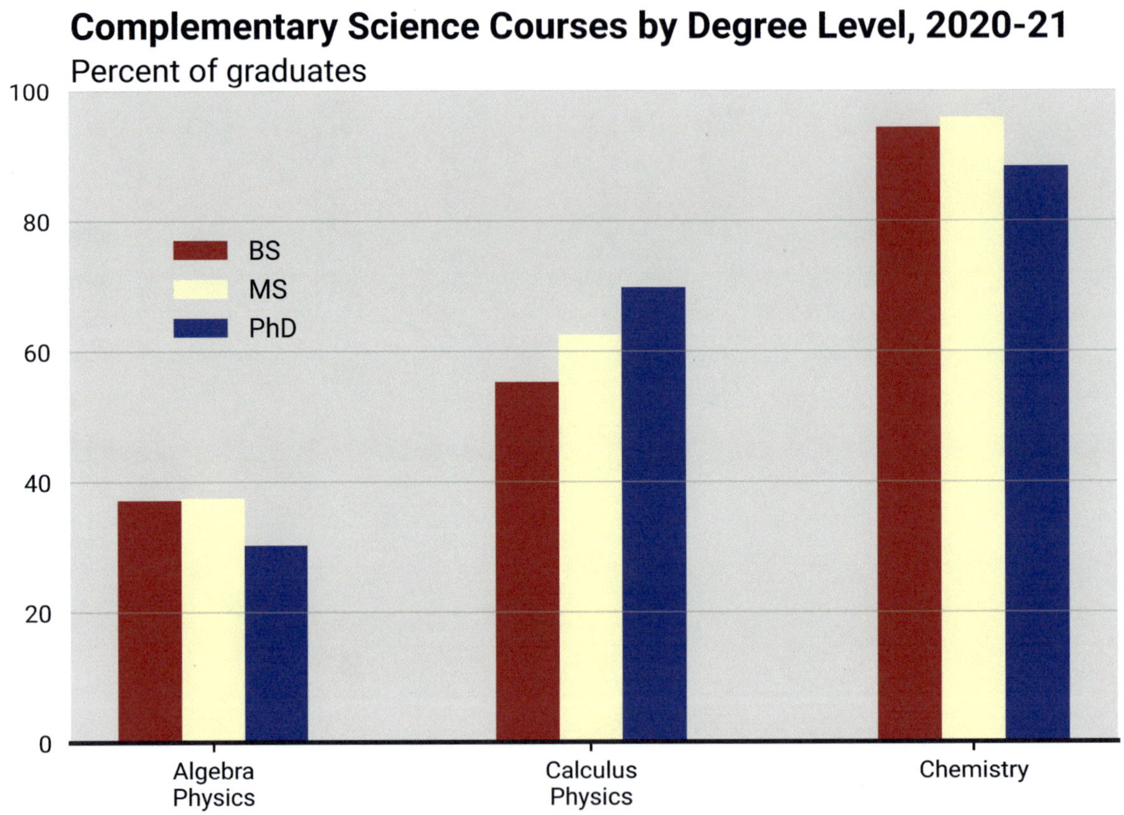

The complementary science courses taken during a graduates' education indicate that chemistry and physics, as topics, are nearly ubiquitous. Interestingly, slightly more than a third of survey respondents indicated that they took an algebra-based physics course, well above the rate of graduates reporting not taking calculus. The nearly 40% of graduates not taking a calculus-based complementary science course could be an indicator of the limited math preparation of many students, calling out the need for either dedicated courses or integration of applied math concepts in geoscience courses to build context and confidence in students' quantitative skills.

2.14 Complementary Sciences Taken, by Gender

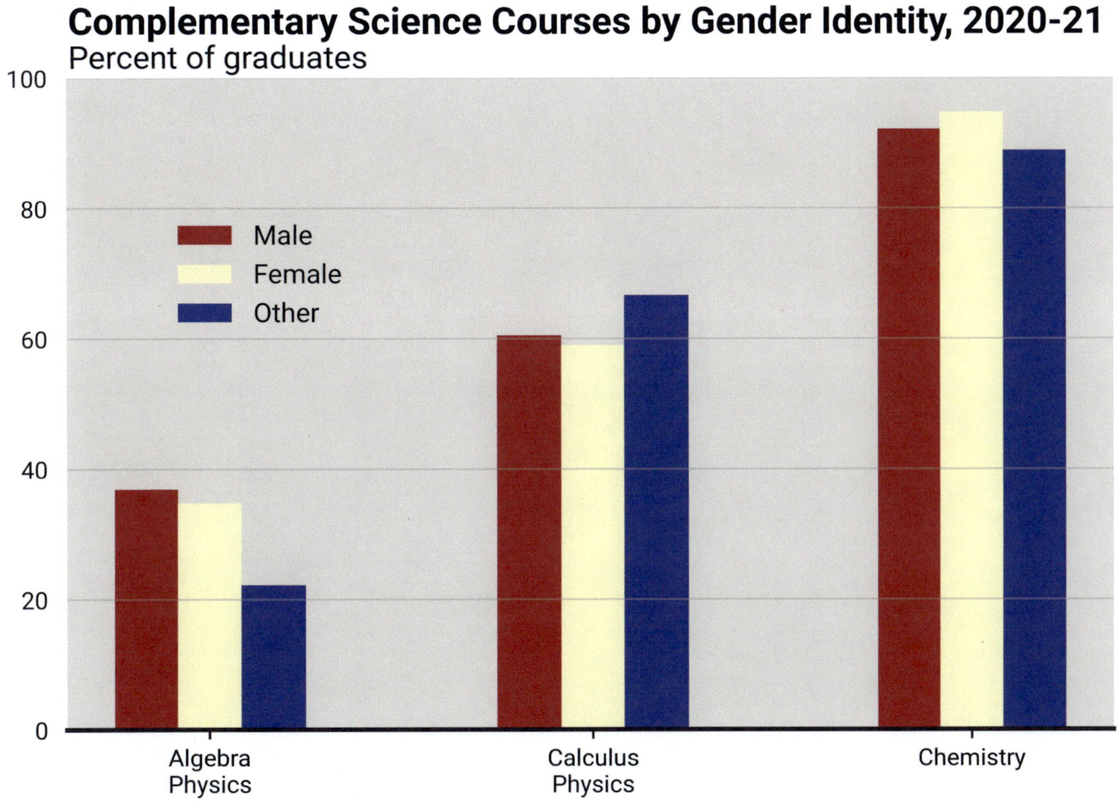

Gender differences across STEM remains a topic of concern and whether there are notable differences in experiences. For complementary science courses, there is no evident variance between male and female participation rates, while the population size of other gender classifications are too small to be significant in their differences.

2.15 Quantitative Skills

Quantitative Skills by Degree Level, 2020-21
Percent of graduates

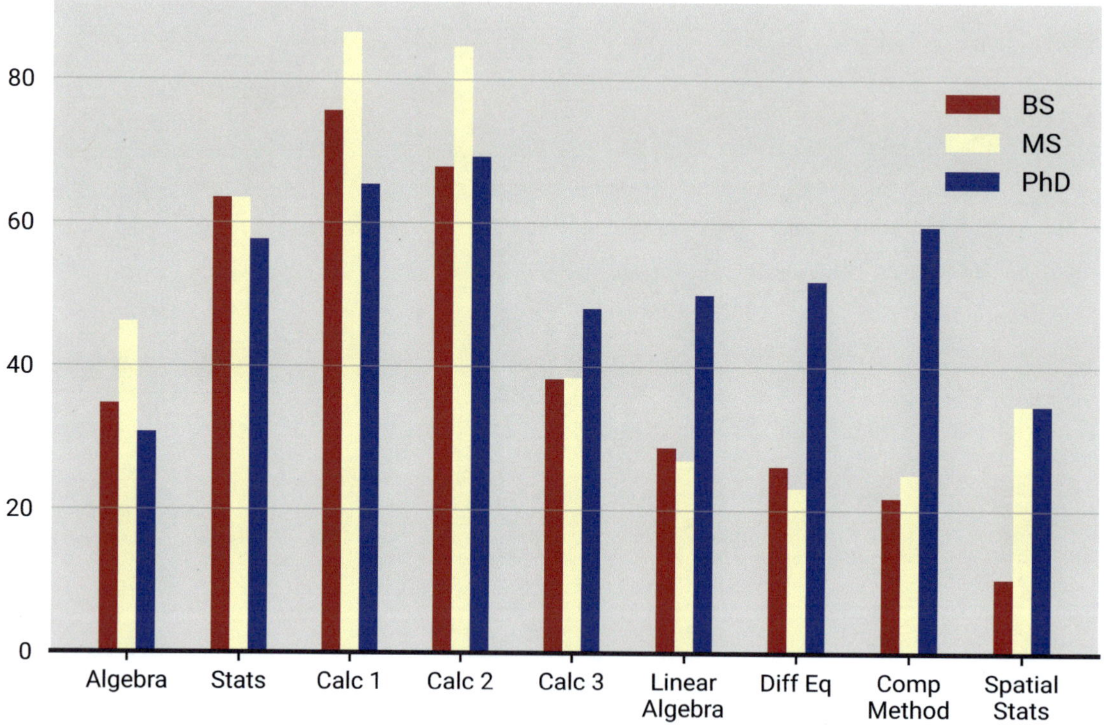

Most bachelor's graduates took Calculus 1 or 2. As in prior years, bachelor's graduates that took any advanced quantitative courses such as linear algebra tended to take most of the advanced quantitative courses.

Master's degree recipients had similar distributions of quantitative skills to bachelor's graduates but with notably higher rates having taken Calculus 1 or 2. Doctoral graduates had substantially more advanced quantitative skills even as there were fewer individuals having taken Calculus 1 or 2, representing distinct cohorts of individuals who came from science backgrounds versus other fields. Statistics remains a particularly weak point for the geosciences, especially at the doctoral level, where most fields would expect all doctoral graduates to have had formal instruction in statistics.

2.16 **Quantitative Skills by Gender**

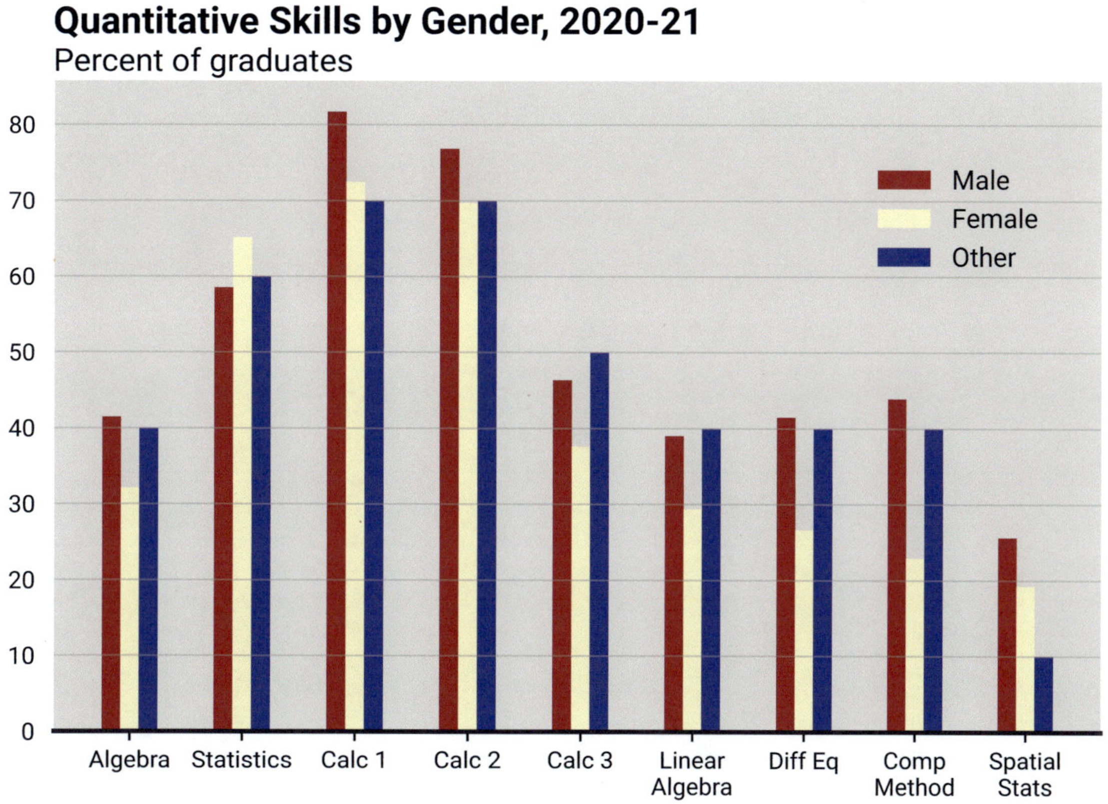

Exposure to various quantitative skills was examined for gender-related trends. Generally, males had greater exposure across all quantitative skills than females, except for statistics. The category for other gender had a very limited population which precluded any significance being assigned to those specific results. What is uncertain is if the observed gender differences are driven by student choice or if specific programs and fields of study which may have different quantitative requirements may themselves have had gender distributions that were substantially different from the overall geoscience population.

2.17 **Doctoral Graduate Exposure to Statistics**

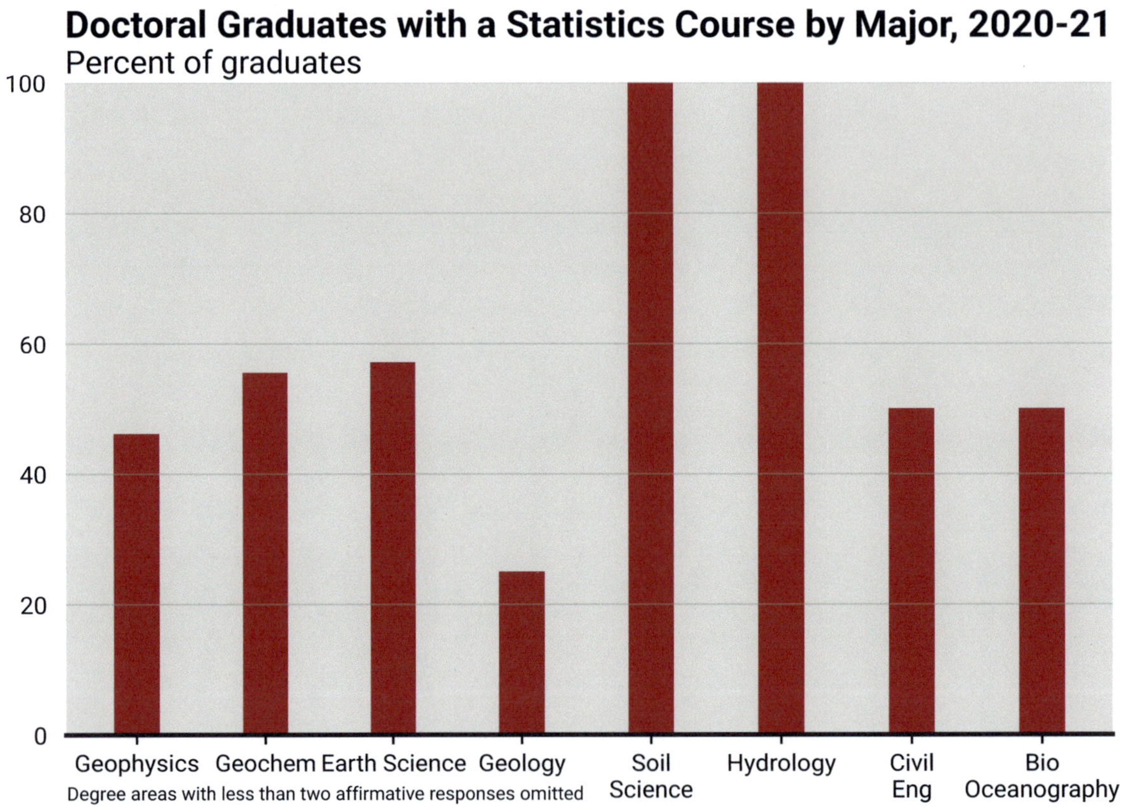

Doctoral graduates have historically had modest exposure for formal statistics courses compared to other STEM fields. The statistics experience of doctoral graduates was examined by field of study. Soil science and hydrology graduates all indicated having taken at least one statistics course, while approximately 50% of graduates from most other fields indicated having taken statistics. However, graduates with geology doctorates, which is one of the larger fields of study at the doctoral level, had weak exposure to formal statistics courses with only about 1 in 4 graduates having taken such a course. Clearly, only a few fields of study are demonstrating a strong requirement for statistics, but the general weak uptake across the three largest fields of study—geophysics, geochemistry, and geology depressed the overall participation rates for geoscience graduates.

2.18 Computational and Data Skills of Bachelor's Graduates

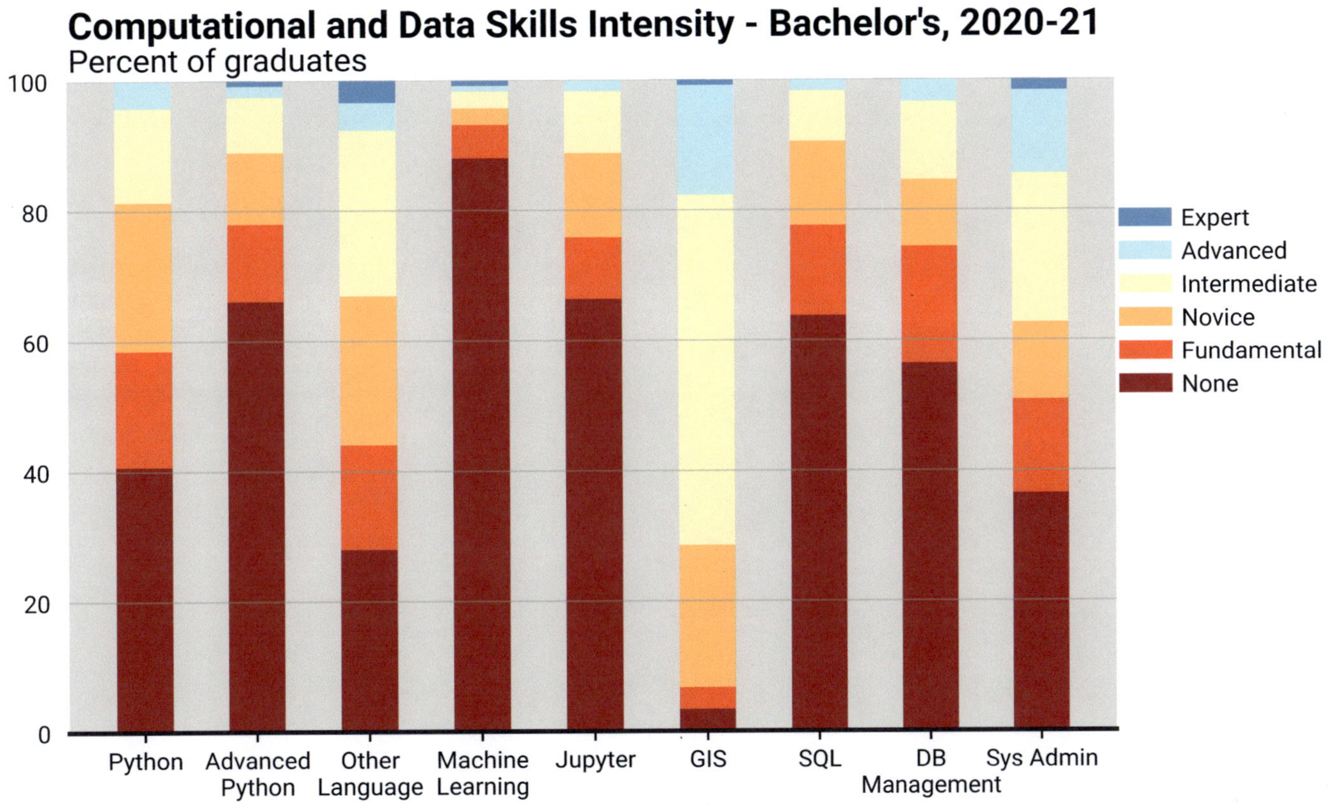

Bachelor's graduates reported limited exposure and competency in most computational and data skills. Only two areas had more than 50% of bachelor's graduates reporting at least a novice expertise—GIS and "Other Programming Languages" which is probably reflective of the use of either R or Java in supporting coursework.

2.19 Computational and Data Skills of Master's Graduates

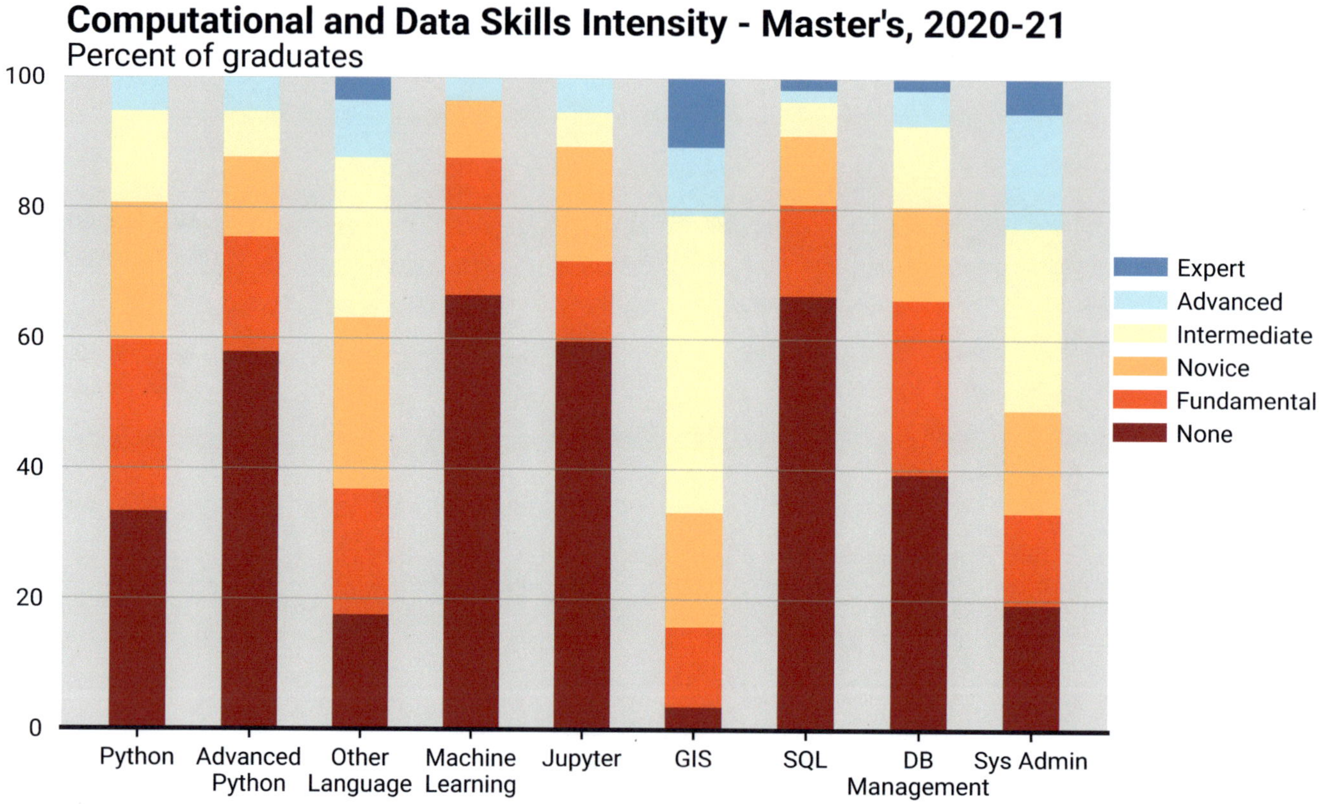

The competency of master's graduates with data and computational skills was marginally stronger than at the bachelor's level, with all skills showing greater levels of proficiency. More than half of master's graduates reported at least novice expertise with GIS, "Other Languages," and System Administration. This is likely reflective of additional courses, as well as—for system administration—the greater level of independent work required at the master's level.

2.20 **Computational and Data Skills of Doctoral Graduates**

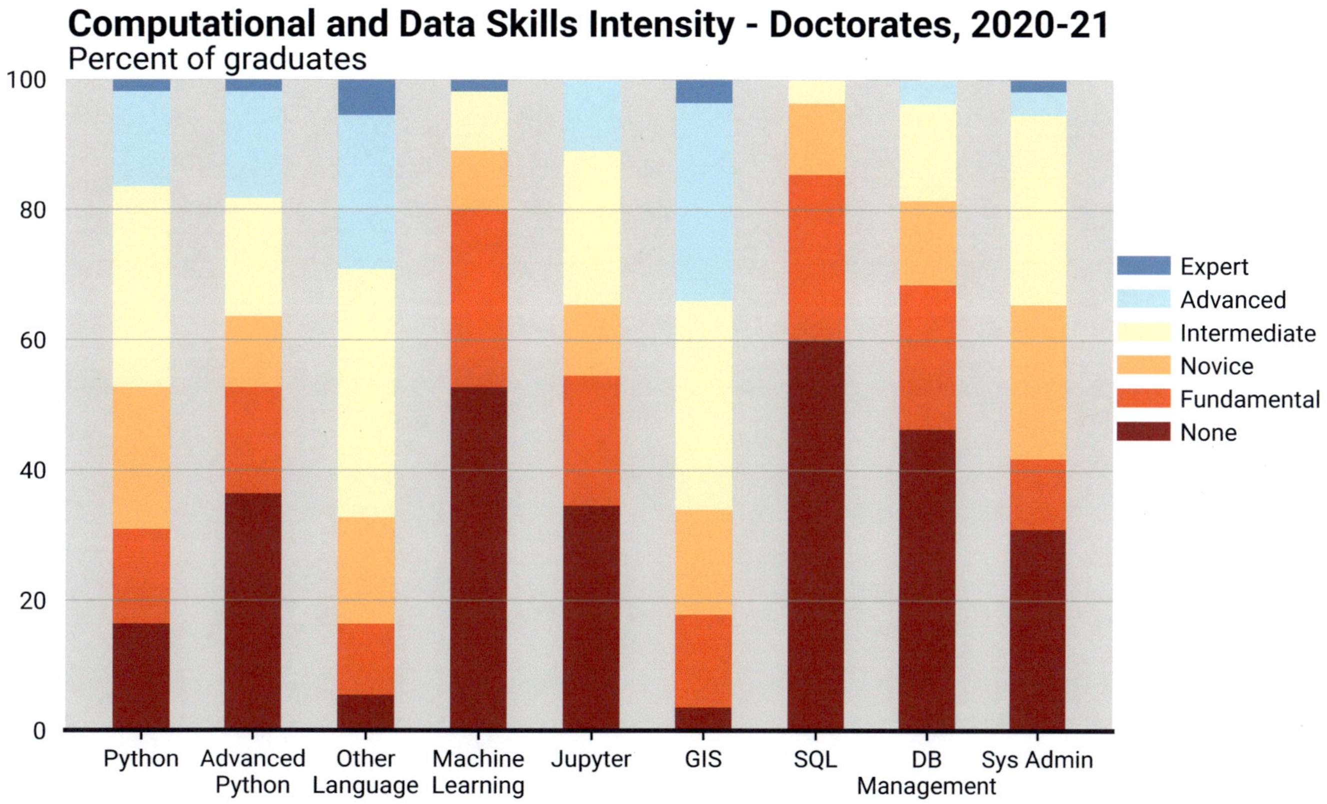

The competency of doctoral graduates with data and computational skills was substantially greater than master's or bachelor's graduates. This is unsurprising given the increased research intensity and work required in doctoral programs. The only areas showing lower levels of competency were in machine learning, SQL, and database management. The lower levels of exposure to machine learning and Jupyter notebooks are likely representative of the rapid evolution and penetration of those tools into the research sphere since the time most of the doctoral graduates would have commenced their research projects.

3. Experiences

Agnieszka Drobniak from AGI's 2016 Life as a Geoscientist contest

3.1 Sources of Financial Support

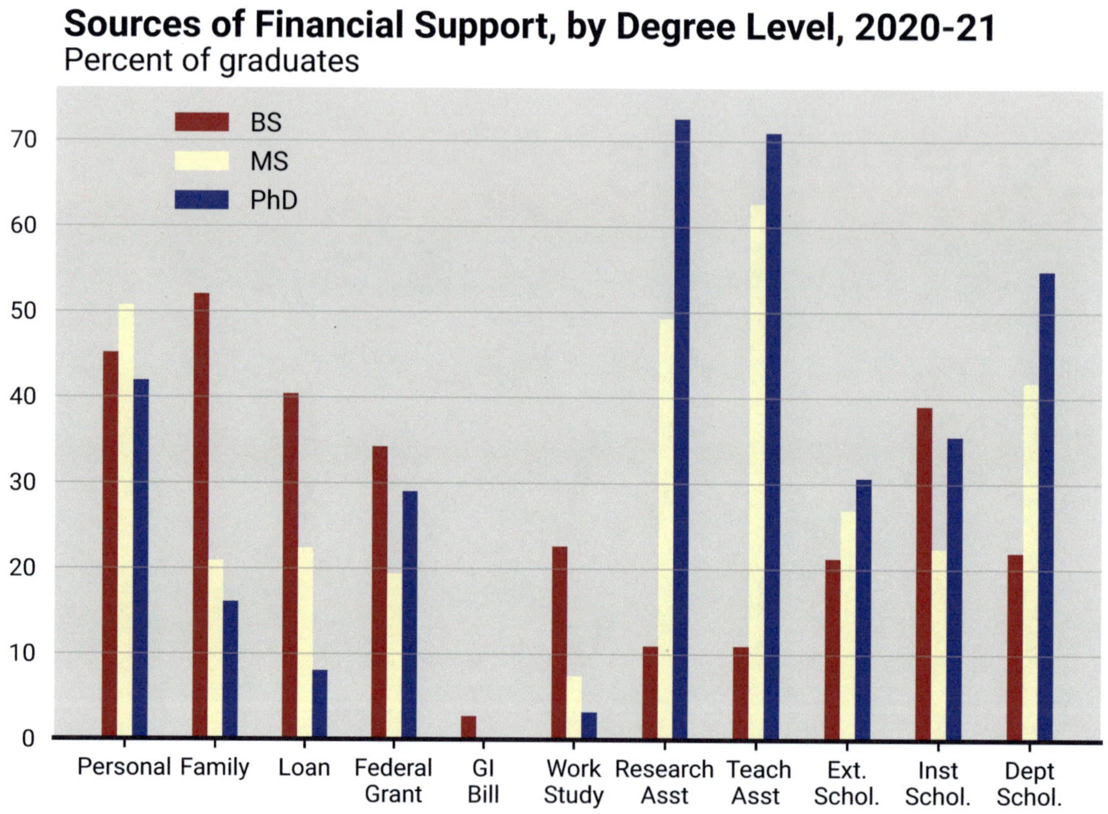

Bachelor's and master's graduates relied on financial outlays by themselves and from their families. Nearly half of all students spent personal funds on their education, and at the undergraduate level over half indicated their families also invested in their education. Loans and federal grants rounded out the support for bachelor's graduates.

Research and teaching assistantships as well as other scholarships were the major source of funding for doctoral graduates and to a lesser degree, master's graduates. The low reliance on work-study is reflective of the socioeconomics of most geoscience students. Also, the minimal use of the GI bill or ROTC scholarships indicates a growth opportunity for the geosciences to tap a new population with the estimated 80,000 students each year being supported through those programs.

3.2 Number of Independent Research Experiences

Number of Research Experiences, 2020-21
Percent of graduates

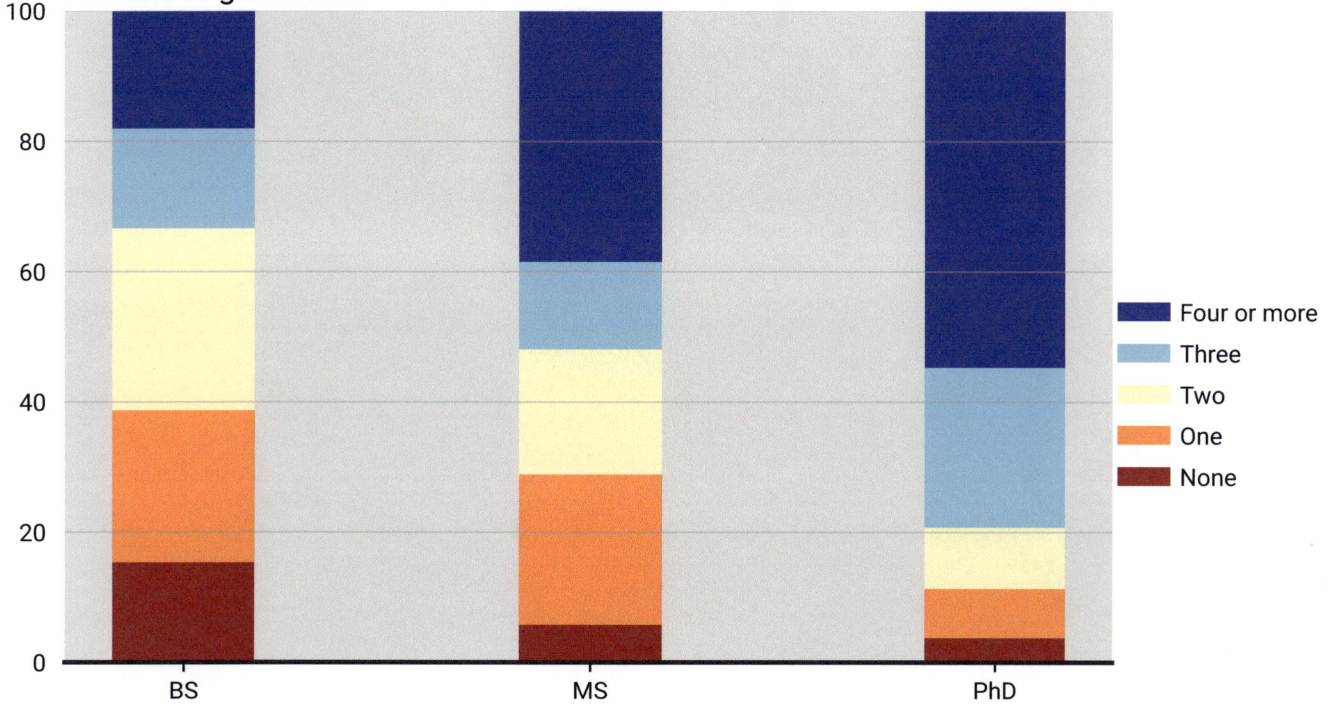

Almost all geoscience graduates had at least one independent research experience, with the rate increasing from 87% to 96% from the bachelor's to doctoral levels. Furthermore, more than 60% of graduates at all levels had two or more independent research experiences, demonstrating the educational and cultural importance of these experiences within geoscience programs. That some doctoral graduates indicated they had no independent research needs further analysis—it may reflect research done as part of a cohort or in cooperation with a faculty member.

3.3 Research with Faculty

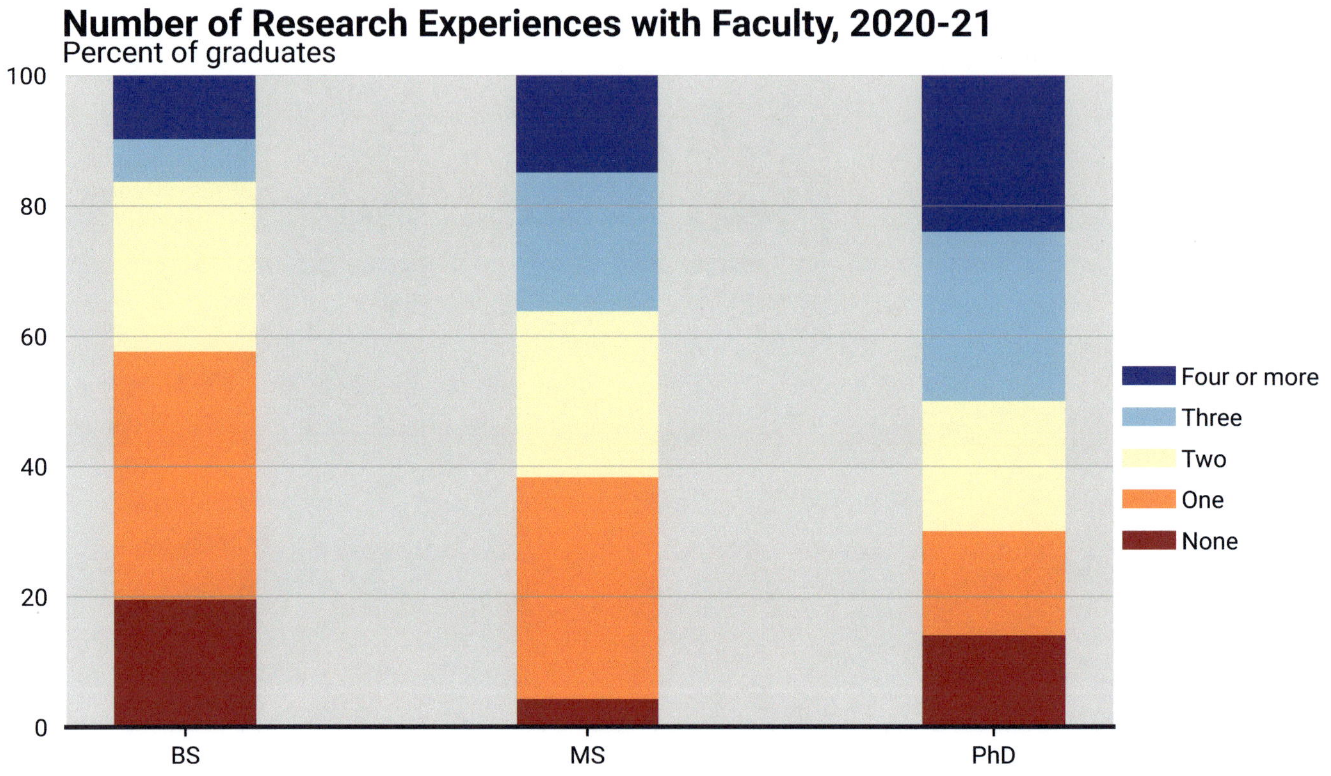

More than 80% of graduates indicated they had at least one research experience collaborating with a faculty member. As expected, the number of individuals with multiple collaborative research experiences grew from approximately 40% to 70% at the doctoral level. Of note is the small percentage of master's graduates indicating they had no research experiences collaborating with faculty. The non-thesis master's graduates would be most likely to not have these research experiences, but this did not appear to be a factor within the survey-respondent population given the nearly ubiquitous research experience with faculty.

3.4 Research Methods Employed

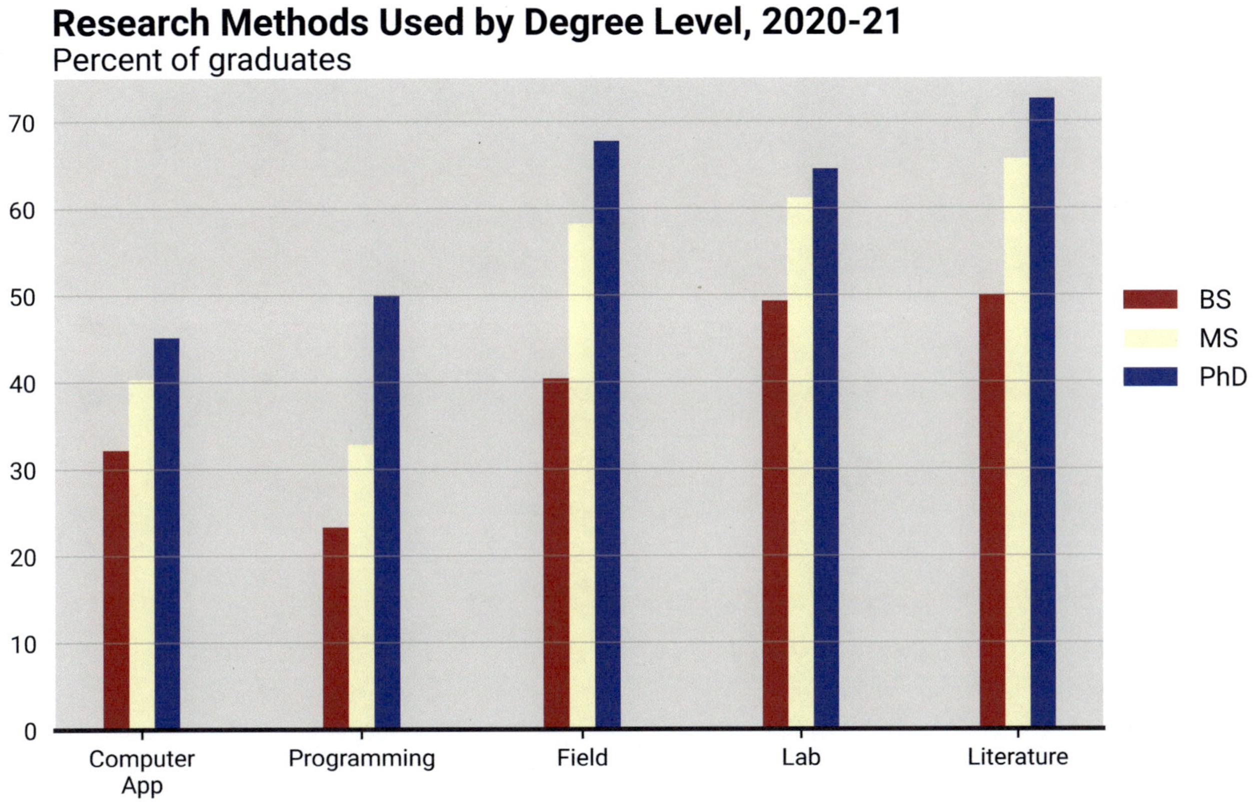

The intensity and diversity of research methods used by graduates unsurprisingly increases by advancing degree level. Except for the use of dedicated computer applications, all methodologies were used by at least half of doctoral graduates, while no single research method was used by more than half of bachelor's graduates.

Literature-based research and laboratory work were the most-used methods across the degree levels, usually taught formally through courses. With greater resources and skills, doctoral graduates employed a more robust and broad set of methodologies. The master's graduates' methodology employment was consistently between that of bachelor's and doctoral graduates.

3.5 Engagement in REUs

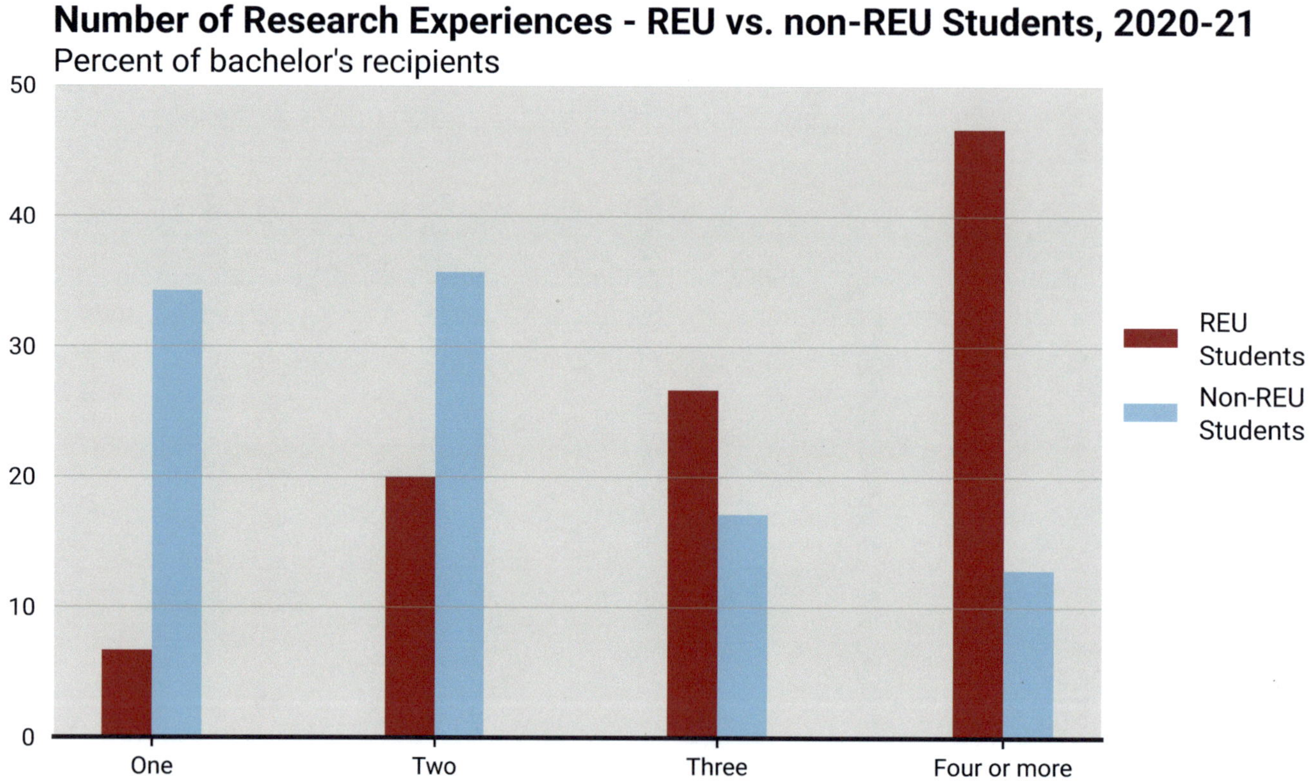

The Research Experiences for Undergraduates program (REU) sponsored by the National Science Foundation, is an important program within the undergraduate geoscience enterprise designed to foster research and science skills for students who otherwise may not have access to these opportunities. Bachelor's graduates who had an REU experience reported a greater total number of research experiences during their education compared to graduates who did not attend REU programs. This trend may be indicative of the developmental benefits both in skills and self-efficacy of undergraduates resulting from engagement with a formal program designed to develop their research skills. Though funded REU opportunities are limited, this result shows positive impacts of early, formal student research experiences on graduate outcomes.

3.6 Joining Geoscience Campus Organizations

Rate of Geoscience Organization/Club Engagement, 2020-21
Percent of graduates

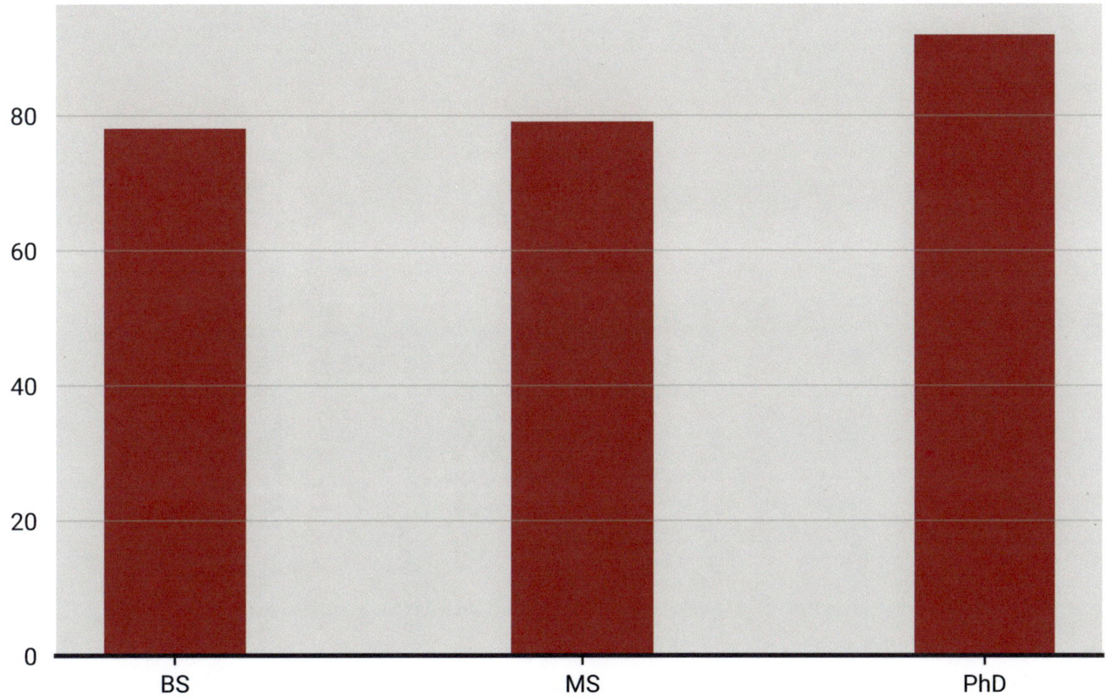

Across all degree levels a substantial majority of graduates participated in an on-campus geoscience organization or club, such as a society student section or geology club. At the doctoral level, nearly all students indicated joining a geoscience organization, which may reflect both membership's advantage when publishing and presenting research and that many doctoral programs are in departments with formal geoscience organizations.

3.7 Finding Internships

Sources to Identify Internships, 2020-21
Percent of graduates

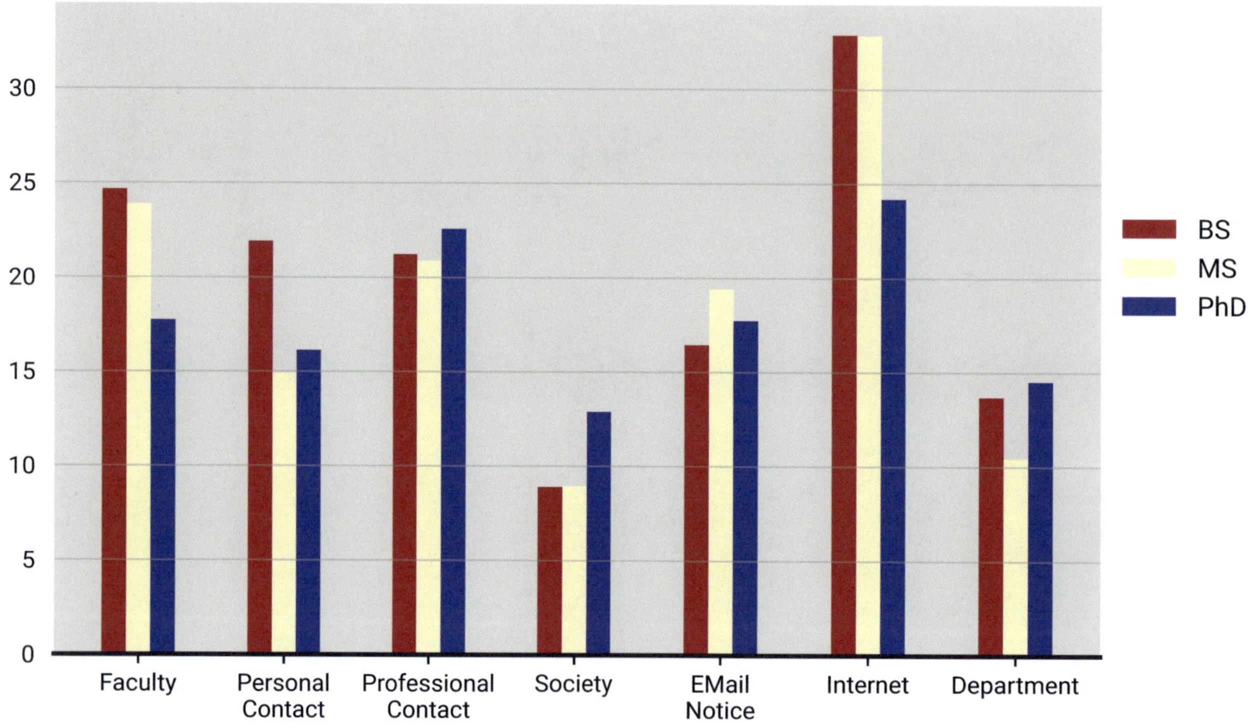

Graduates reported using a spectrum of methods to find internship opportunities, with no single approach used by more than 35% of graduates. The top methods employed by undergraduates were looking on the internet and networking with faculty, friends, and professionals. Methods employed by master's graduates were substantially similar to bachelor's with less reliance on the use of personal contacts. Doctoral graduates often utilized professional contacts but used geoscience societies and departments more than the other degree levels, reflecting the different phase of their professional development.

3.8 Internship Applications

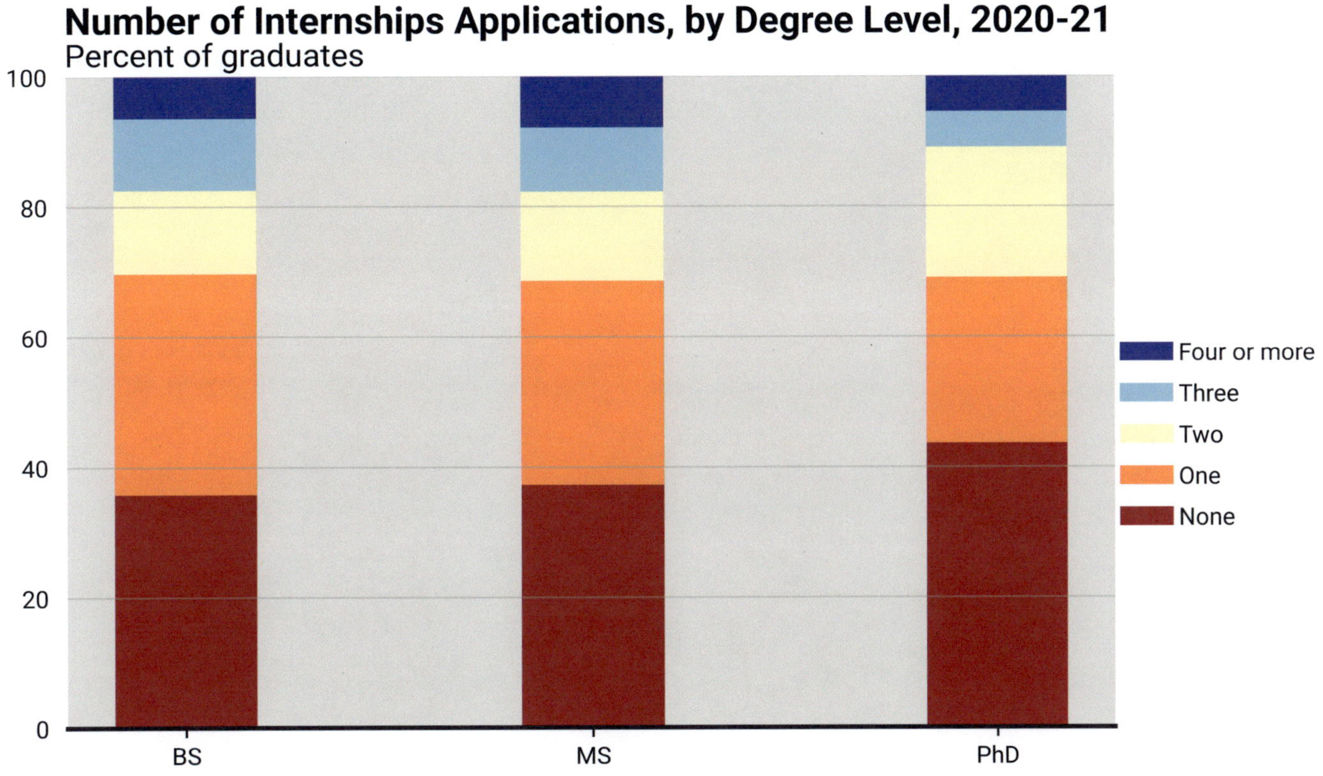

Across all degree levels, about 60% of all graduates applied to at least one internship during their studies. The proportion of students applying to multiple internships was also consistent across the various degree levels, indicating either structural awareness of opportunities (through departments or faculty) or that a consistent fraction of graduates viewed internships as a key part of their education.

3.9 Number of Internships Held

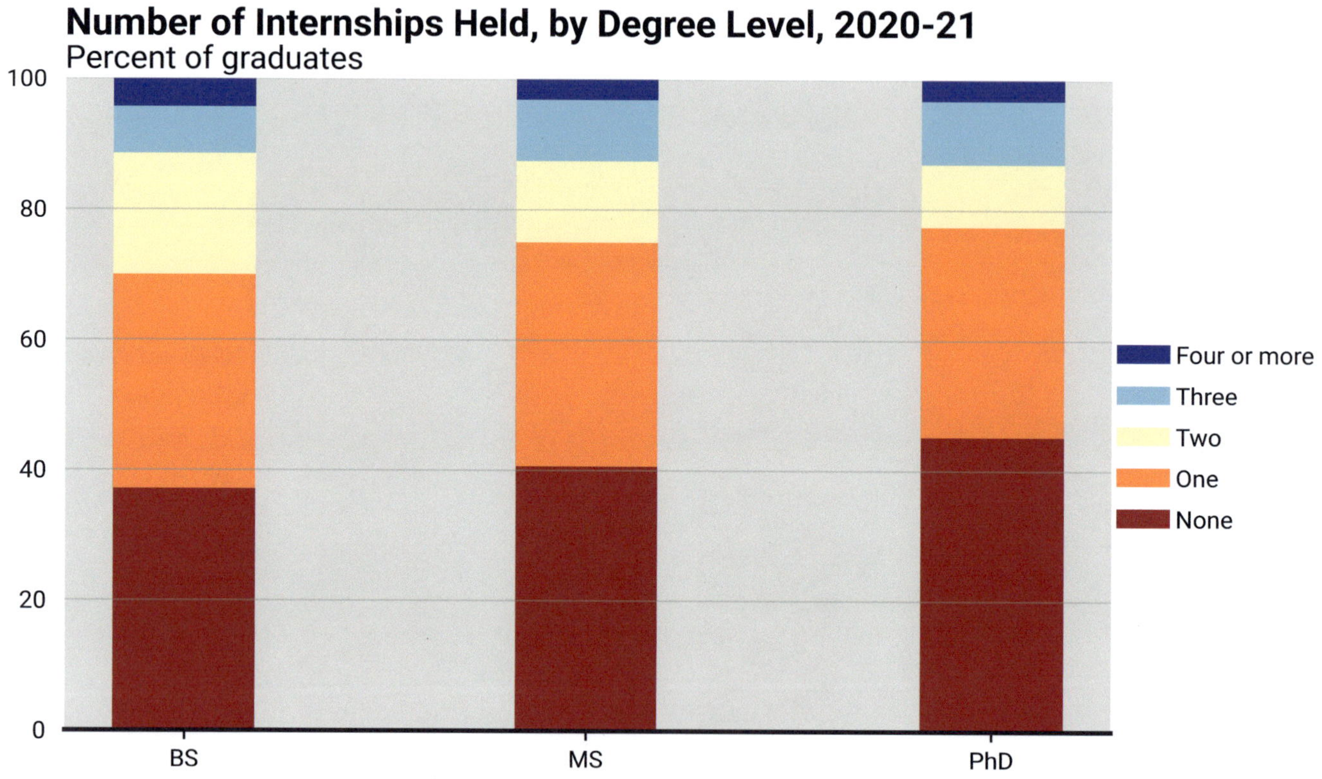

Internships are the major source for career experience and development for geoscience students. Across all degree levels approximately 60% of students held at least one internship during their studies, and 25% had two or more internships. Why some graduates did not hold any internships is unclear—it is unknown if this is from choice because of other commitments like research or from inability to be hired, and whether there were underlying factors for this such as location, time, or skill deficiencies.

3.10 **Success Rates for Internship Applications**

Rate of Success in Securing an Internship, 2020-21

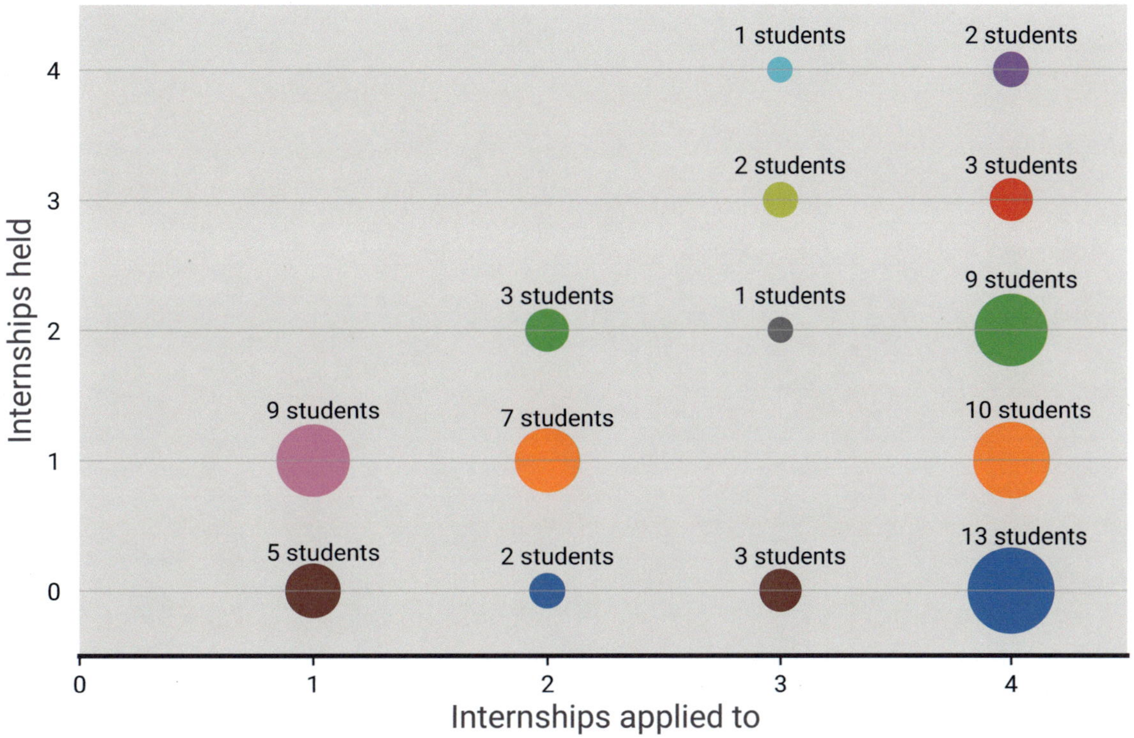

About two-thirds of graduates who applied for an internship secured an opportunity, with 37% of applications leading to an offer. When looking at the matrix of success rates for securing an internship, two specific points arise. First, one student held more internships than applied, as they were given an unsolicited additional offer. Also, 13 graduates reported applying to four internships but were unable to secure any positions. The factors that drove this lack of success is unclear, whether it was related to location, skill level, or alignment to education level at the time of application.

3.11 **Perceived Value of Internships**

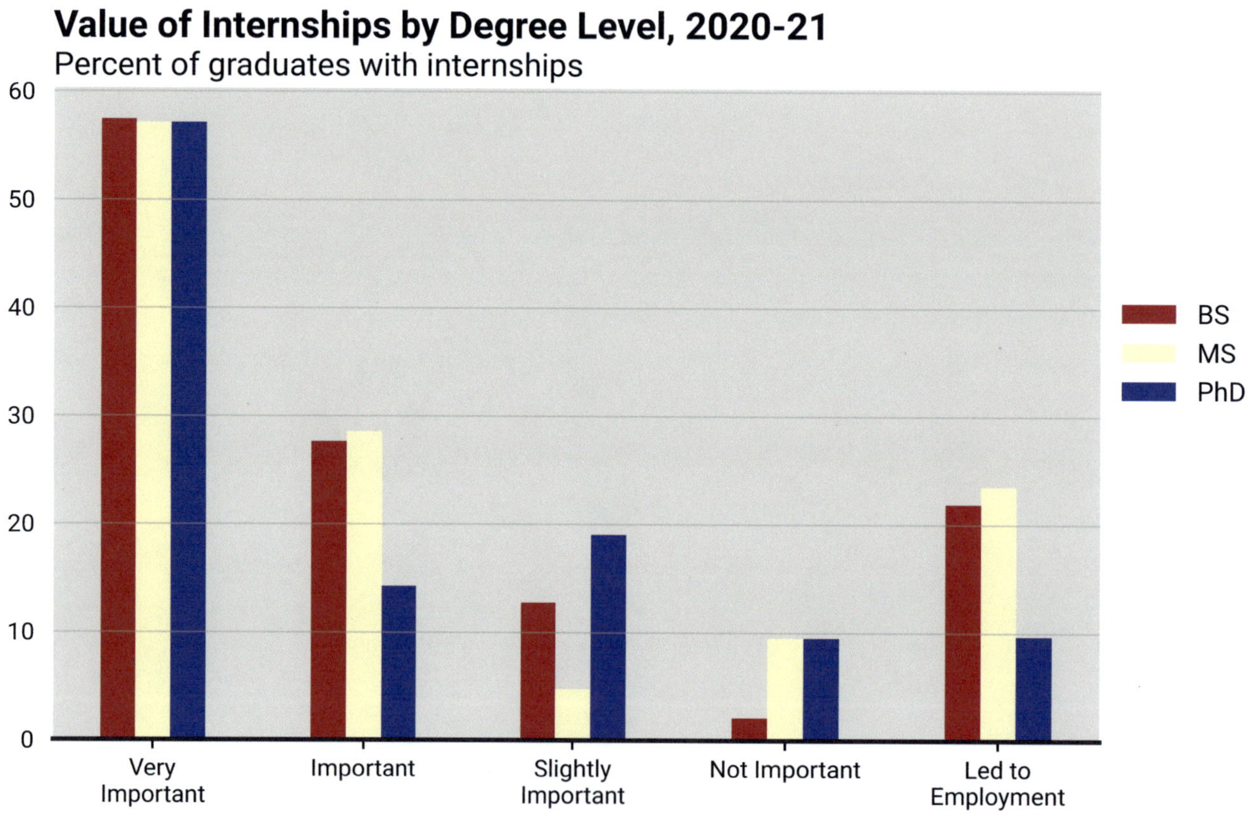

Across all degree levels, most students who held an internship rated it as very important or important to their professional development. Only graduate students were likely to report either minor or no importance from internships, likely because of a misalignment with their career or research goals. Internships remain an important gateway to post-graduation employment with over 20% of bachelor's and master's students who held an internship securing employment with the same organization after graduation.

3.12 **Job-seeking Strategies**

Job Seeking Resources by Degree Level, 2020-21
Percent of graduates

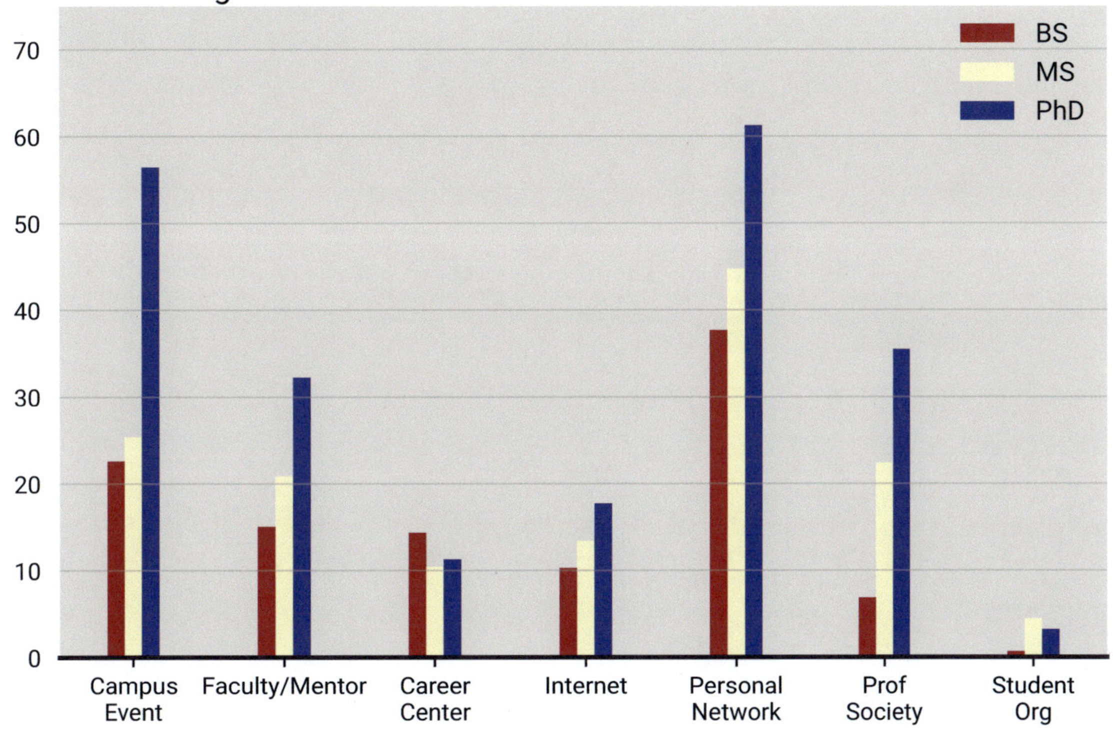

Doctoral graduates used a wider range of job-seeking resources, reflecting the added exposure to networking options. Across all degree levels, personal networks were the singular most important job-seeking tool, aligning with the general trend that most geoscience jobs are found through networking. Doctoral graduates also relied heavily on professional societies and campus events to seek employment. Of note, student organizations, the internet, and career centers were minor resources for job searches across all levels, which is particularly curious given the importance of the internet for finding internships.

4. Outcomes

Amy Edwards, for AGI's 2014 Life as a Geoscientist Contest

4.1 Graduate School Application Status

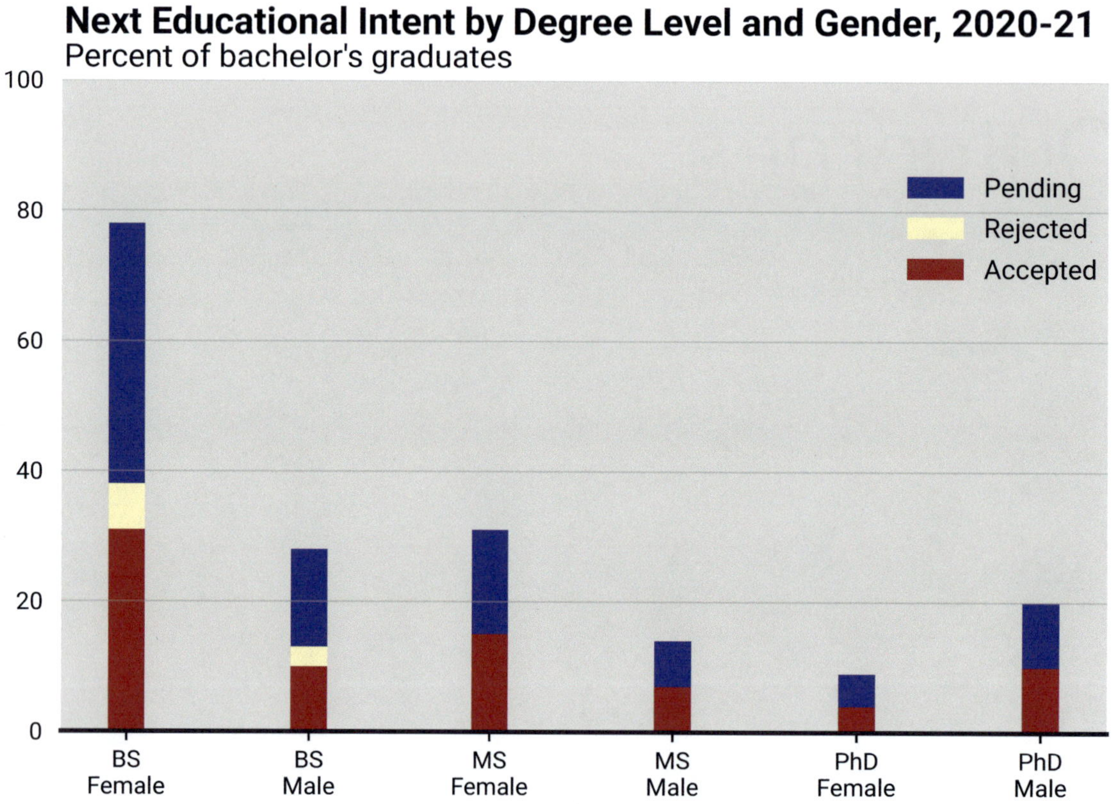

Female bachelor's graduates were more likely to continue their education than any other group of graduates. Across all gender/degree levels cohorts, about half of applicants had received an acceptance decision on their graduate school applications by the time of the survey (late Spring 2021). Only bachelor's graduates reported rejections. Women were twice as likely as men at the master's level to apply to additional graduate programs. Interestingly, approximately 20% of the doctoral recipients intended to continue their education, but in this case more men than women were seeking further education.

4.2 Intended Field of Study for Next Degree

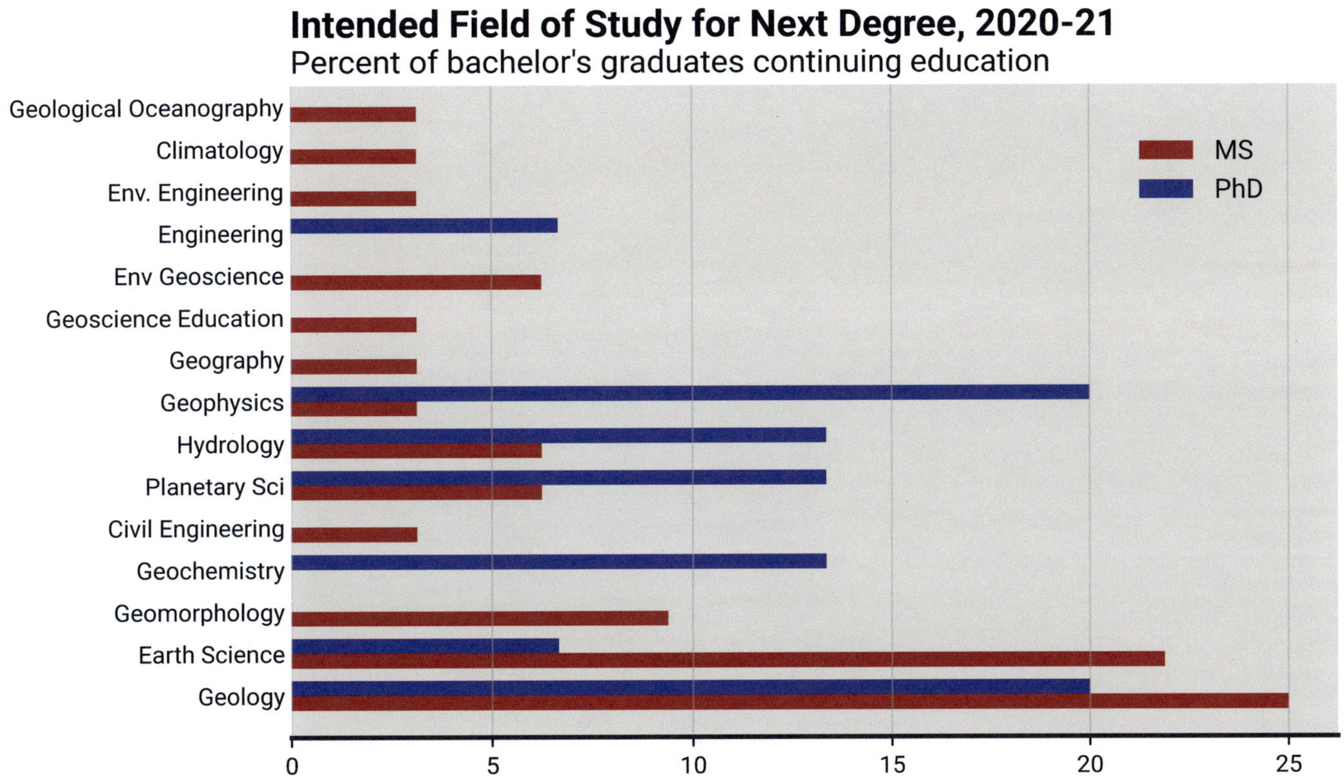

Bachelor' graduates showed a greater diversity of target fields of study for master's programs than for doctoral programs. For those graduates intending to pursue a doctorate, the fields of geophysics, geology, geochemistry, hydrology, and planetary science were the intended majors. Geology and earth science, along with applied fields such as geomorphology, geoscience education, environmental geology, and engineering-related areas dominated the choices of master's-intending graduates indicating the varied motivations for a master's degree.

4.3 Degree Sequences of Geoscience Graduates

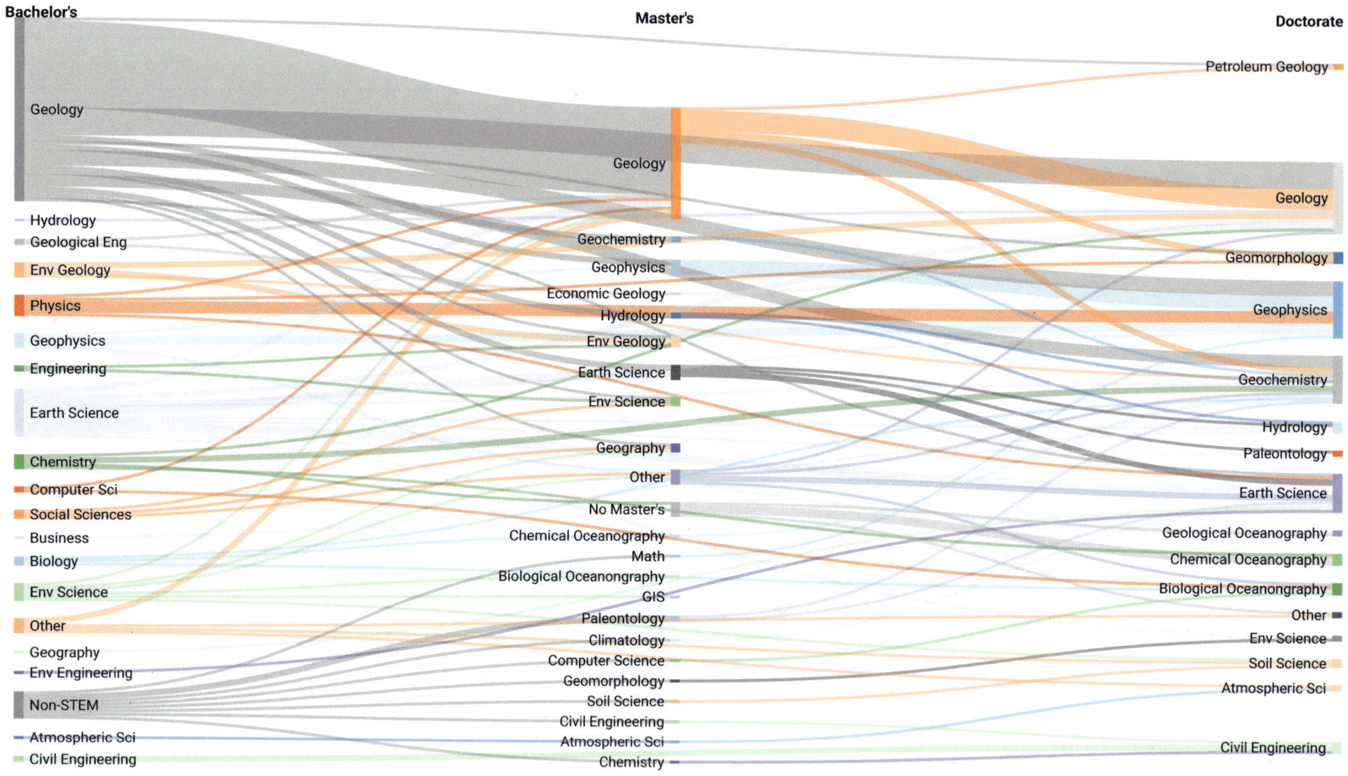

The sequencing of fields of study across degree levels was examined by asking all graduate degree recipients about their prior degrees, even if it was not in the geosciences.

Geology is the most dominant field for bachelor's degrees but declines at each subsequent degree level. Complementary sciences such as physics and chemistry feed directly into several geoscience graduate areas, especially geophysics, geochemistry, and oceanography. The input from complementary sciences and non-STEM fields is particularly important at the doctoral level and we see this evidenced in doctoral responses related to traditional geoscience activities like field work.

4.4 Seeking Professional Credentials

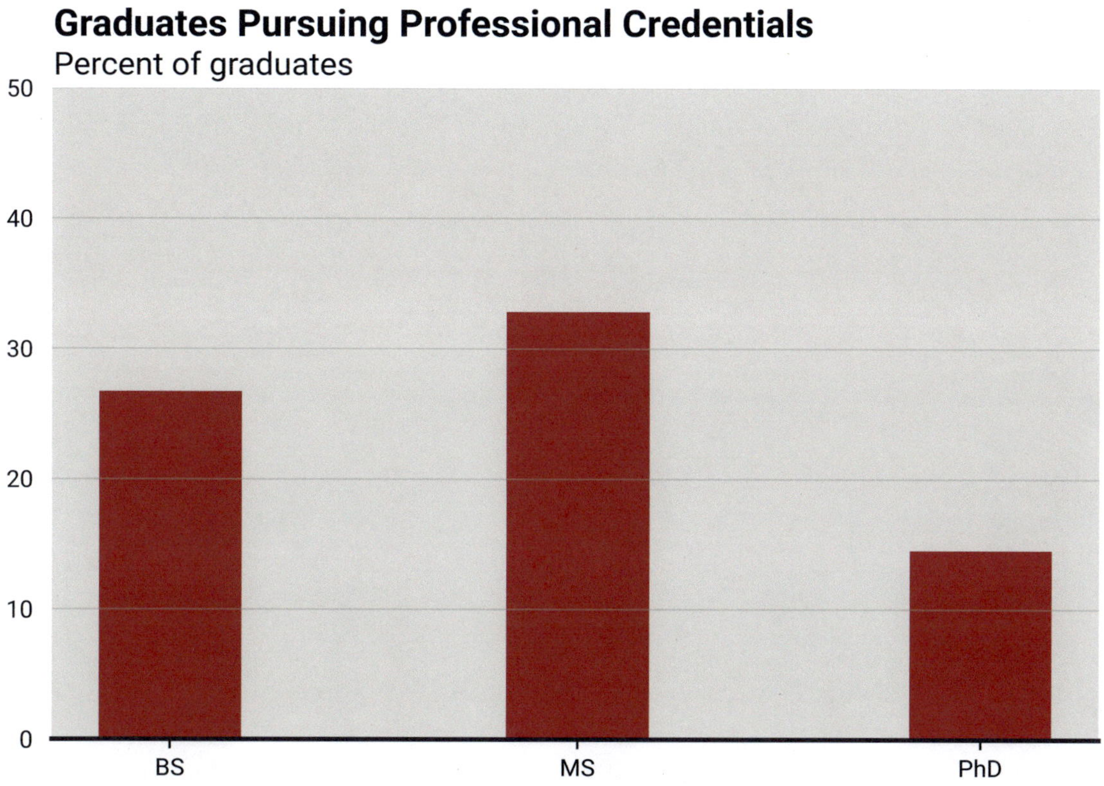

Fewer than half of all graduates indicated that they were pursuing a professional credential, with master's graduates most likely to be seeking a credential. The low rate of seeking credentials is counter to current trends towards increasing employer demand for additional professional credentials as jobs shift towards professional services. Whether for state licensures, OSHA or similar certifications, deficiencies in formal professional credentials are a challenge for recent graduates in securing employment.

4.5 Credentials Being Pursued

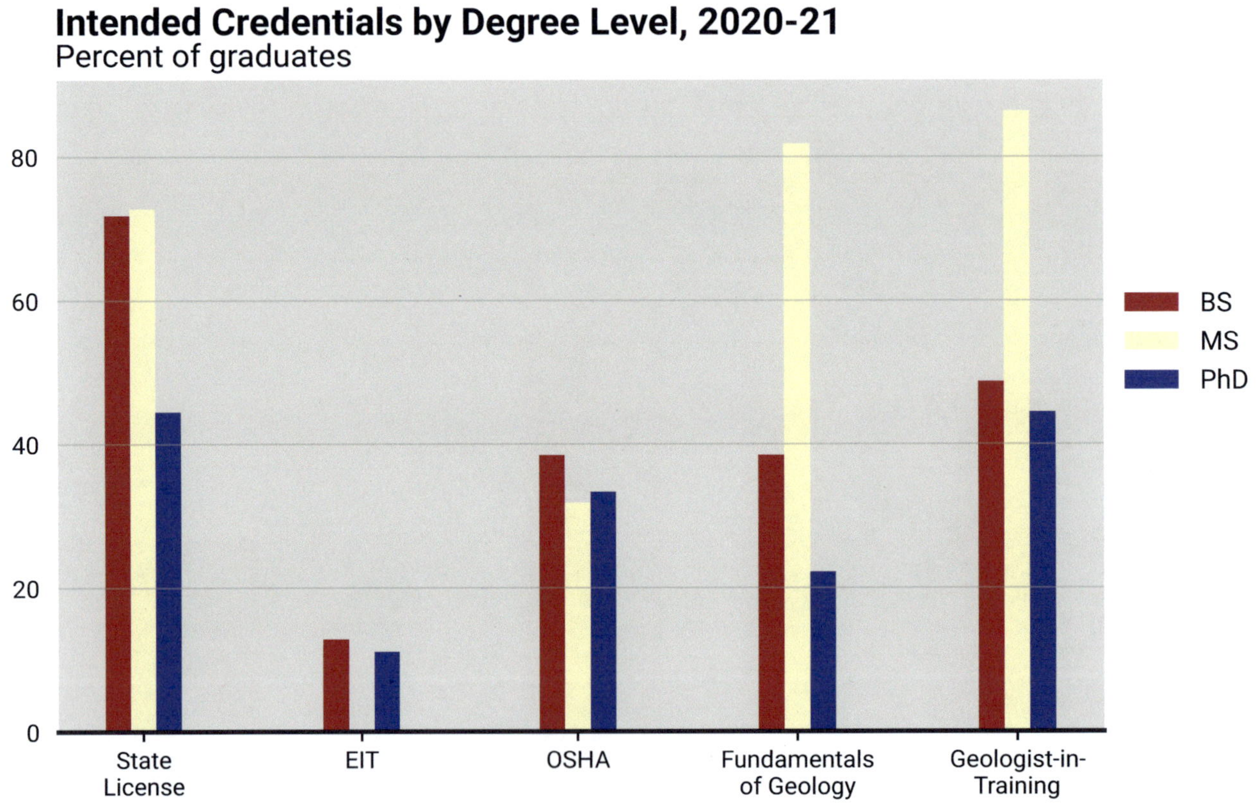

Of graduates indicating they were seeking a professional credential, state licensure was the ultimate goal for most of those respondents, either directly or through taking the related ASBOG exams or Geologist in Training (GIT) qualification. Curiously, intent to take the Fundamentals of Geology exam and securing GIT was most prevalent among master's graduates which is different from recommendations for candidates to take the exam immediately following graduation from a bachelor's degree program.

About 30% of all graduates seeking credentials were looking to get OSHA certifications, such as HAZWOPER, and this indicates due diligence by these individuals for specific employment-related requirements, especially in the professional services sector.

4.6 Employed at Graduation

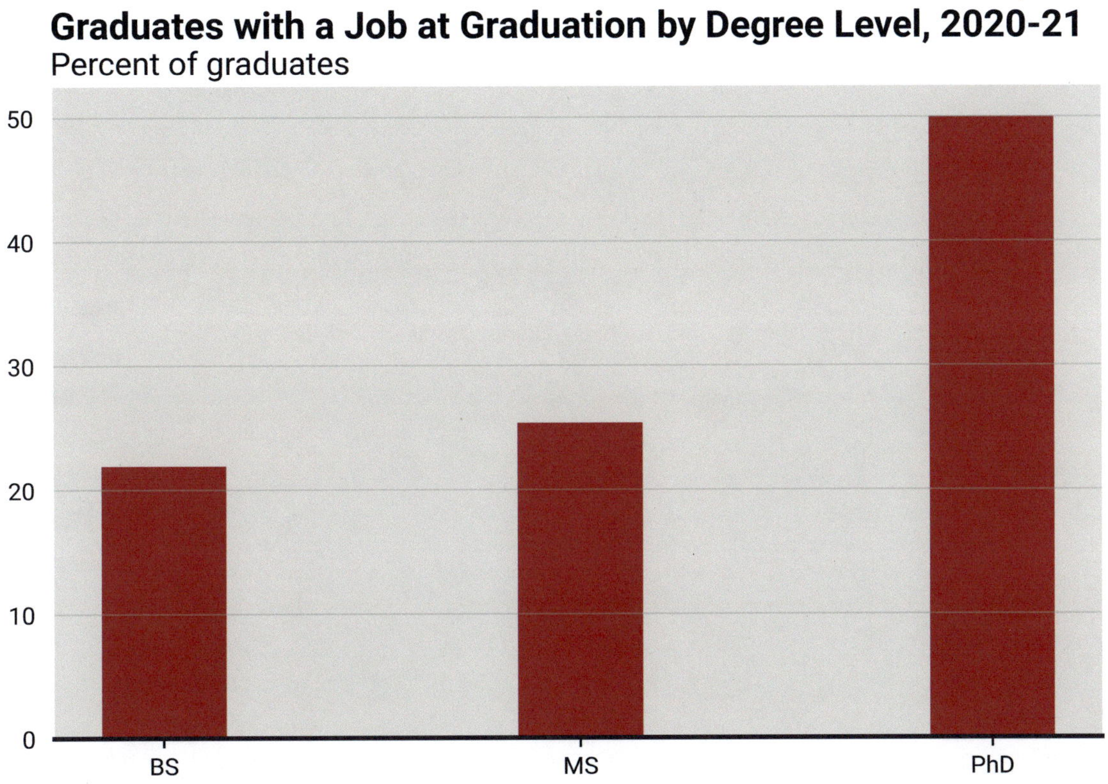

Approximately 20% of bachelor's and master's graduates secured employment by the time of graduation and 50% of doctoral graduates were employed by the completion of their studies. Doctoral graduates had the advantage of possessing a more developed professional network and access to employment with long lead-times tied to academic terms.

Even with the pandemic, these rates of employment at graduation are identical to long-term trends for each degree level and do not represent any negative ongoing trend of employability of graduates. Traditionally, most graduates secure professional employment within three to six months of the end of their studies.

4.7 Employment Status at Graduation

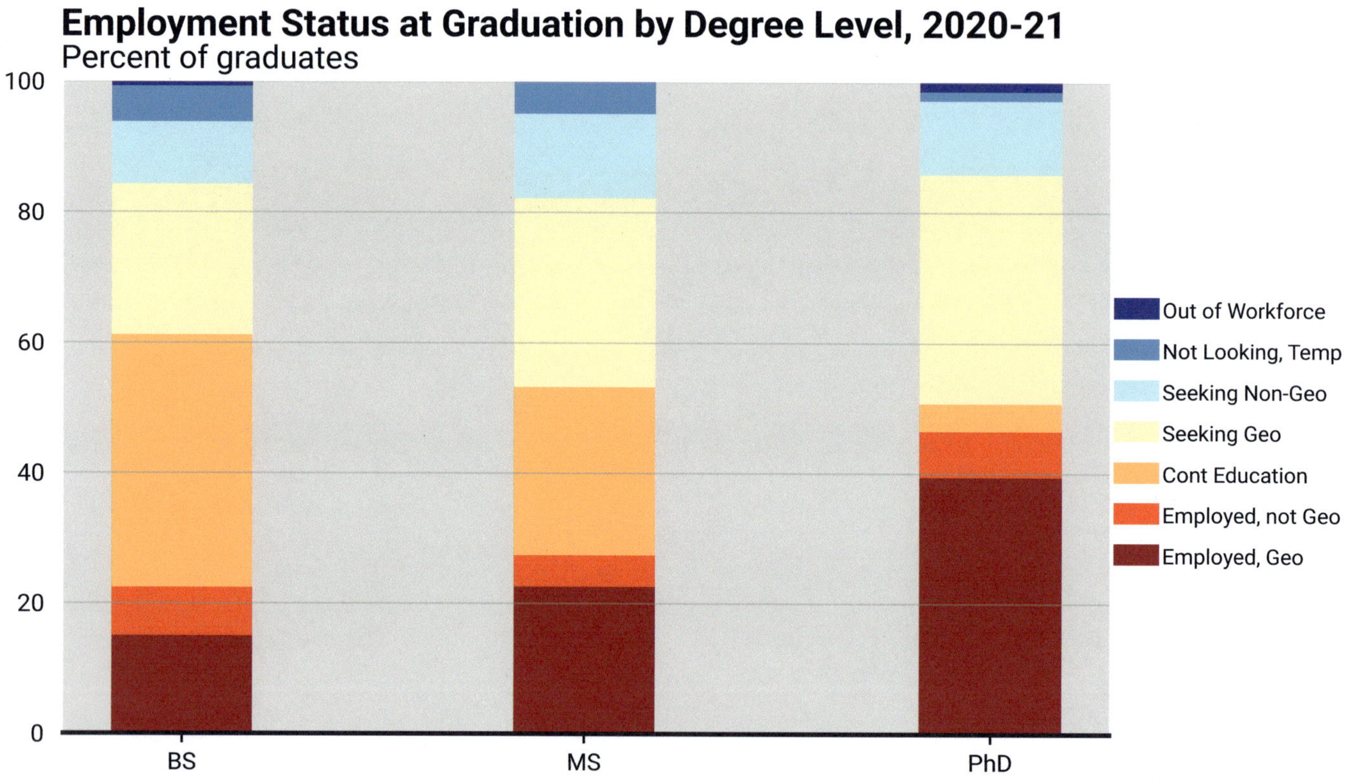

At graduation, usually 20–25% of bachelor's and master's graduates and 50% of doctoral graduates have secured permanent employment. Even with the COVID-19 pandemic, these trends remained consistent in 2020–21. For all degree levels, most graduates were employed or seeking employment in the geosciences. Approximately 20% of all graduates were employed or seeking employment outside of the geosciences, reflecting a positive outlook for professional geoscience prospects.

4.8 Employment Status at Graduation by Gender

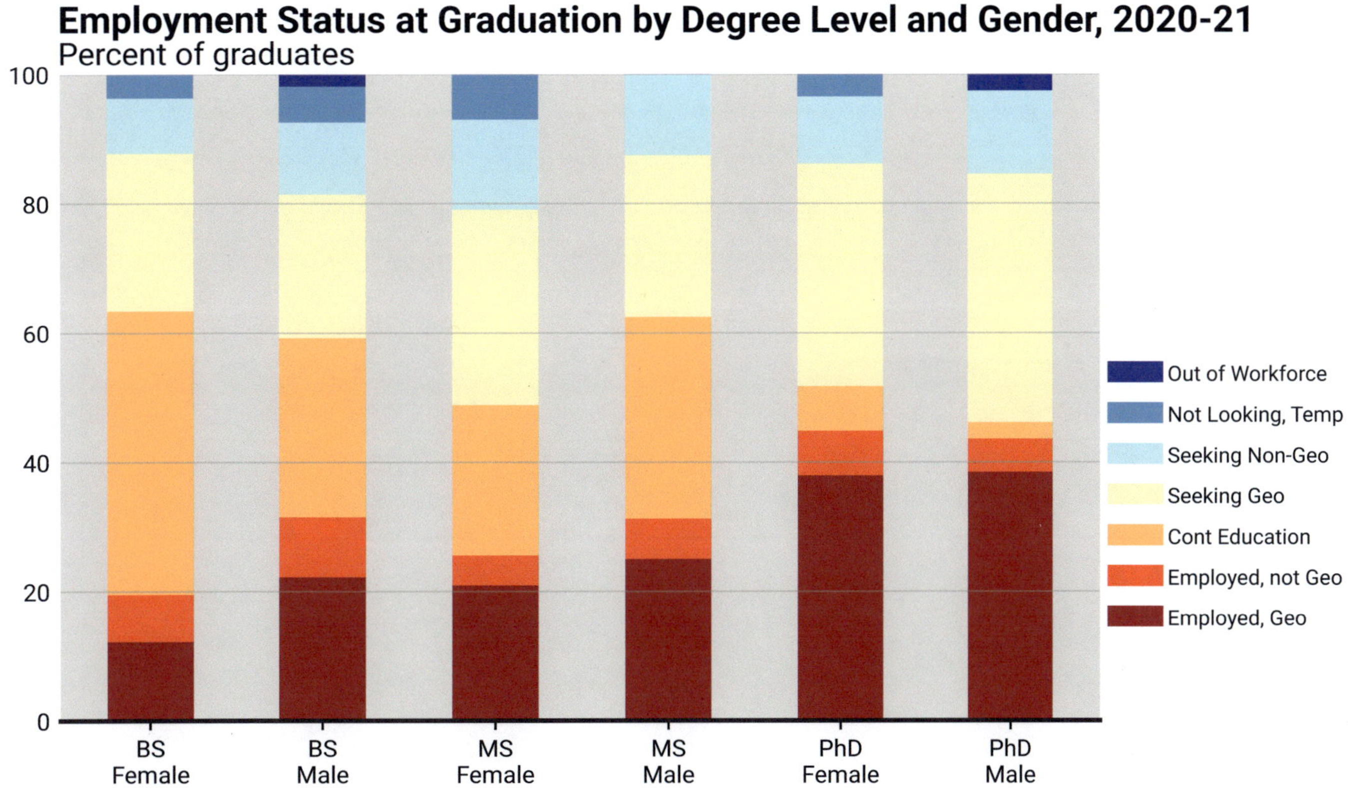

Male bachelor's and master's graduates tended to have secured employment at graduation at a higher rate than did females at the same degree levels. However, at the doctoral level, employment rates were at parity. Female bachelor's graduates were continuing their education or seeking employment in the geosciences at a much higher rate than their male counterparts.

4.9 Where Graduates Sought Jobs

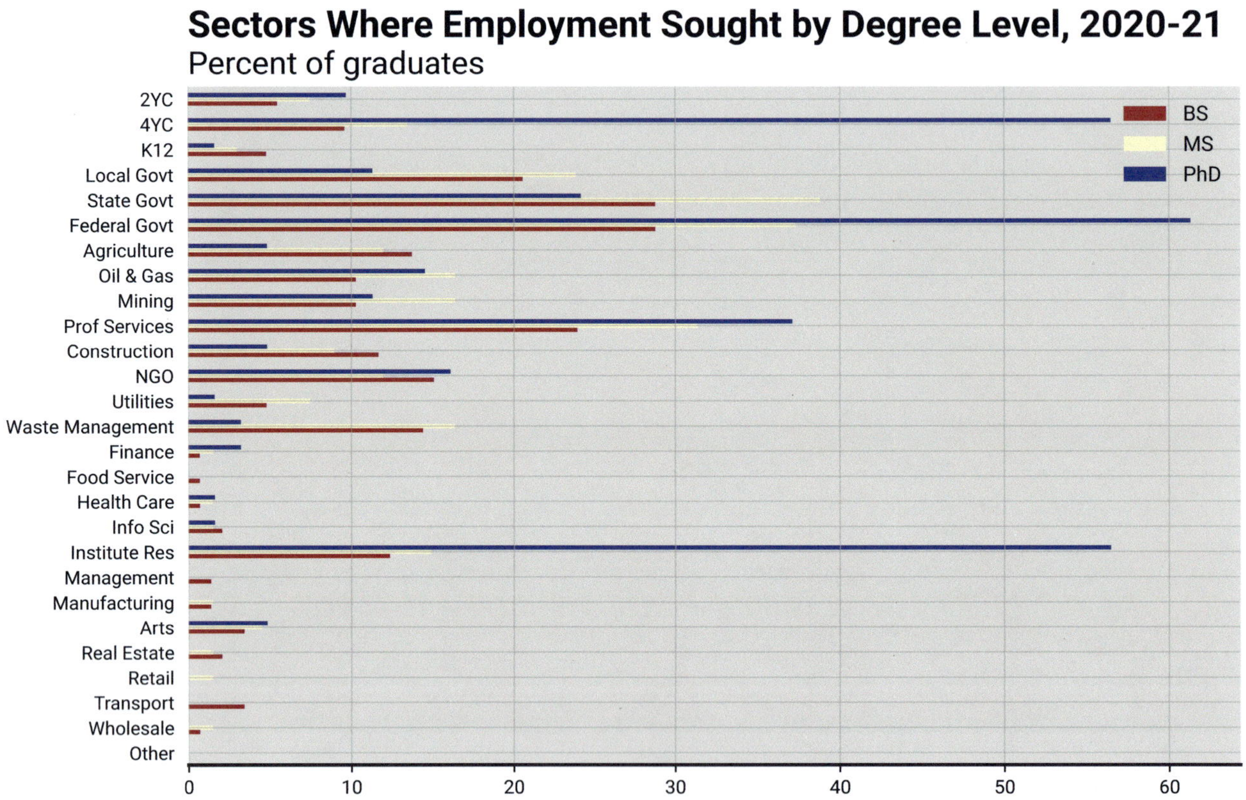

Graduates across all levels sought employment in traditional geoscience areas such as academia, government, resources, and professional services. Bachelor's and master's graduates primarily sought government and professional-services jobs, while doctoral graduates focused on education, federal agencies, and research institutes. Interestingly, doctoral graduates were also the most likely to seek employment in professional services, which is a distinct increase from prior years and resulted in a substantial increase in doctoral employment in that sector. Also evident is the wide range of sectors to which geoscience graduates applied for positions, which is the norm historically with geoscience having healthy skill transferability and many "singleton" geoscience positions within non-geoscience organizations.

4.10 Employment Sectors of Bachelor's Graduates

Employment Sectors of Bachelor's Graduates, 2020-21

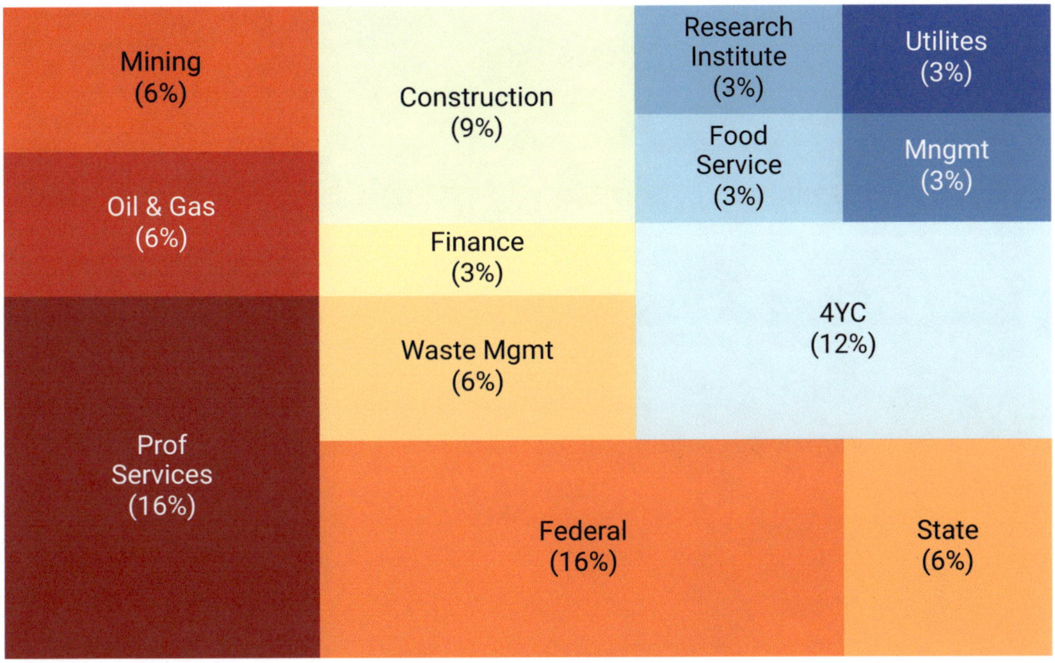

Bachelor's graduates employed at graduation found jobs across a wide spectrum of industries. Professional services and the federal government were the largest employing industries. Mining and oil/gas also were hiring, but the specific allocation of this demand is unclear as to whether these were direct hires into those industries or if graduates were in professional services firms that worked on resource-related projects. Employment of bachelor's graduates at the federal level is often focused on technical field and laboratory geoscience positions, similar to technician jobs available to bachelor's graduates in four-year colleges.

4.11 Employment Sectors of Master's Graduates

Employment Sectors of Master's Graduates, 2020-21

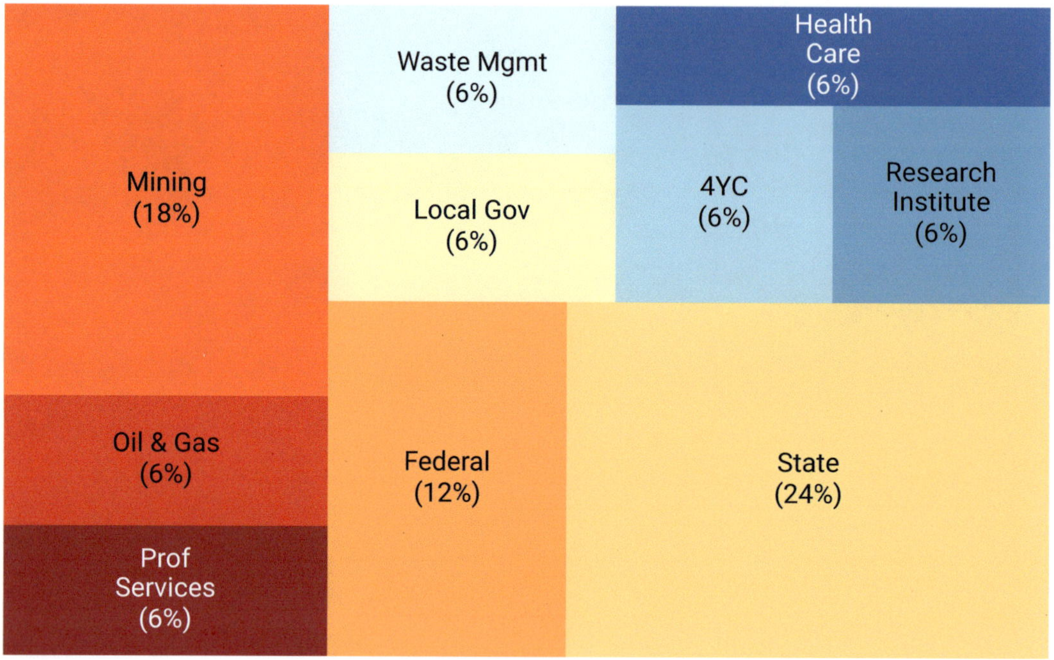

Mining and state government were the two sectors hiring the most master's graduates. The mining sector has been reporting significant demand for new geoscience talent for several years. The increased state government hiring is interesting during the pandemic—it is unclear if hiring is replacement for retiring staff or reflects that many state government budgets were very healthy at the end of 2020. Interestingly, 6% of respondents indicated going into healthcare, which historically has had a minor but steady consumption of geoscience graduates from fields such as geophysics working in medical technology.

4.12 Employment Sectors of Doctoral Graduates

Employment Sectors of Doctoral Graduates, 2020-21

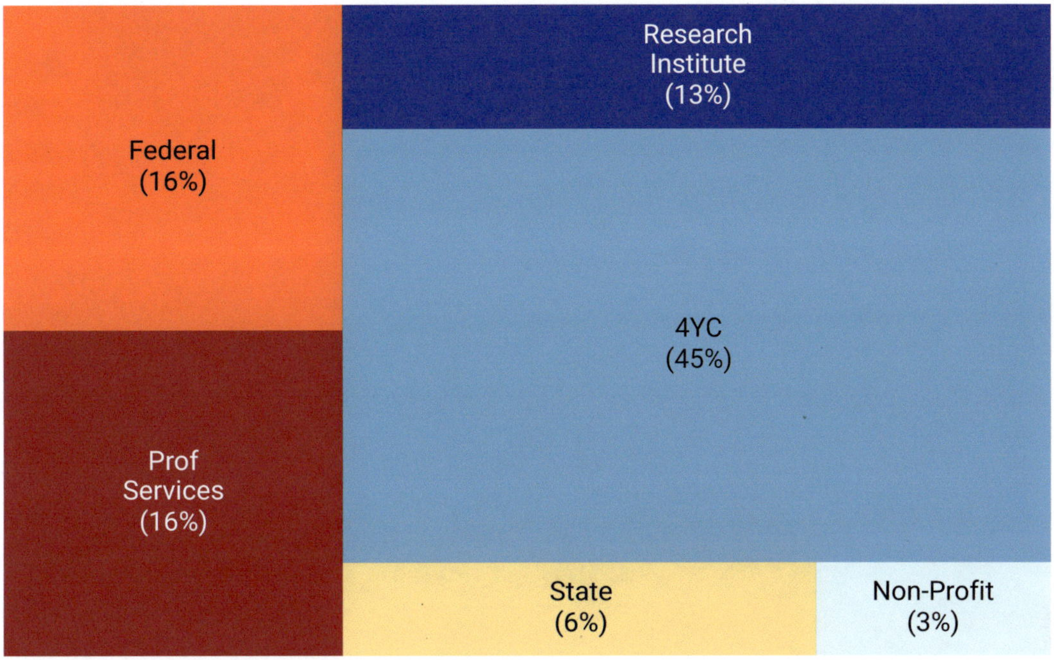

Doctoral graduates historically have found employment within universities and research institutes, and this trend continued for the 2020–21 geoscience doctorates. The most novel development was that 16% of doctoral graduates were employed by the professional-services sector, which historically has not actively hired Ph.Ds. This change represents an increased demand for talent with demonstrable skills by the professional-services sector. Graduate programs will need to consider the impact of the widening skill differential between master's and doctoral graduates as a potential barrier to master's-level employment in the future.

4.13 Employment Trends of Bachelor's Graduates

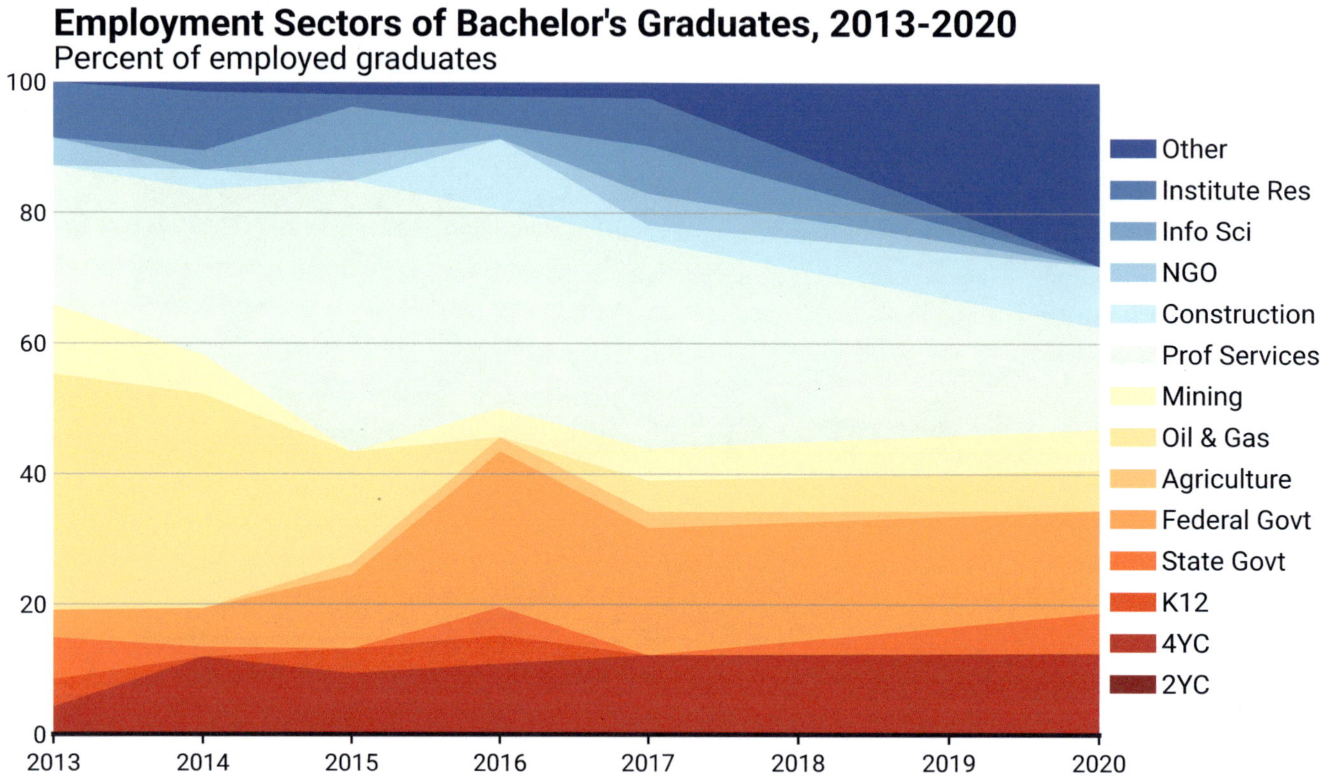

The seven-year trend of bachelor's graduate employment shows some shifts in hiring sectors, with a rapid increase in other sectors consuming about 25% of 2020-21 graduates. Except for professional services, most sectoral hiring has remained consistent, such as the employment of bachelor's graduates as technicians in four-year colleges, universities, and federal agencies. The reduction in some of the more niche sectors such as information science and non-profits, but growth in construction, is likely driven by the pandemic with changing work patterns and needs.

4.14 Employment Trends of Master's Graduates

Employment Sectors of Master's Graduates, 2013-2020
Percent of employed graduates

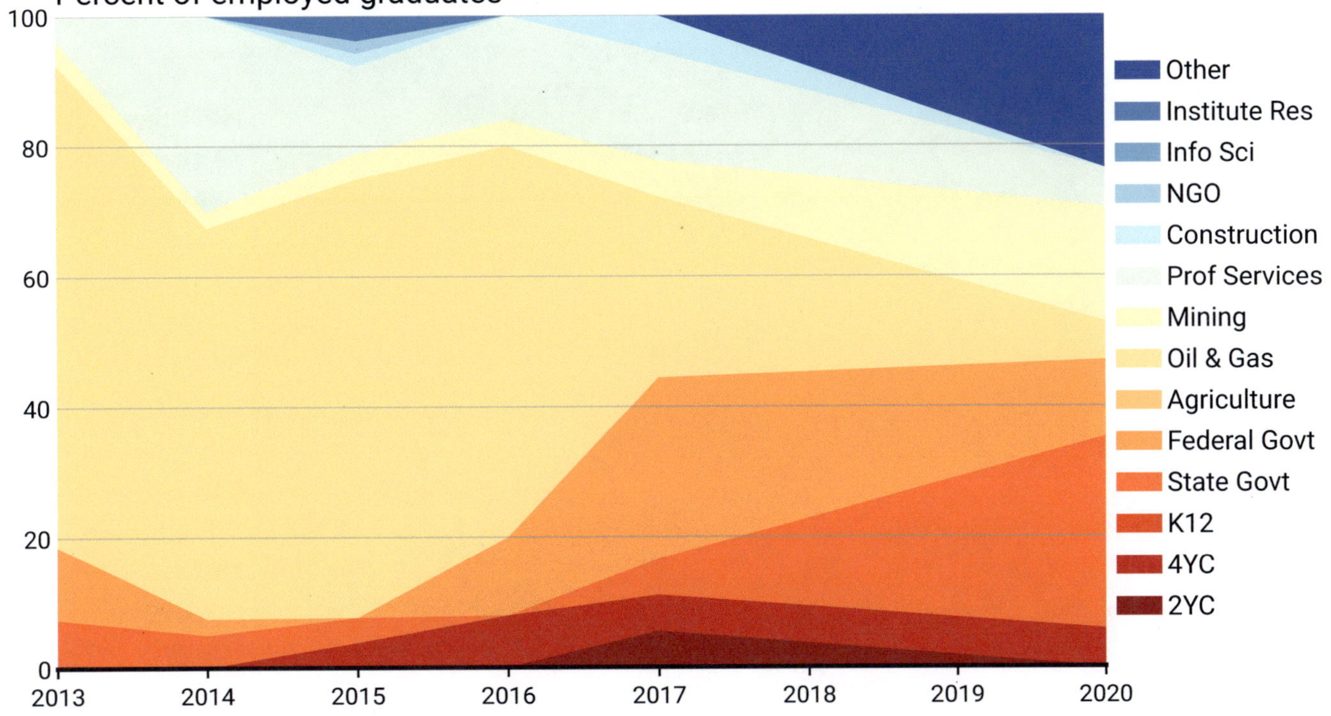

Employment trends at the master's level is defined by the collapse of hiring in the oil and gas sector which had represented between 50 and 70% of all jobs held by new master's graduates earlier in the decade. In ensuing years, with no measurable increase in overall unemployment of master's graduates, these losses have been offset by increases in hiring by state government agencies and the mining sector. Government hiring at the master's level is focused on replacement of retiring technical professional positions, while hiring in the mining sector reflects the economic expansion in that sector.

4.15 Employment Trends of Doctoral Graduates

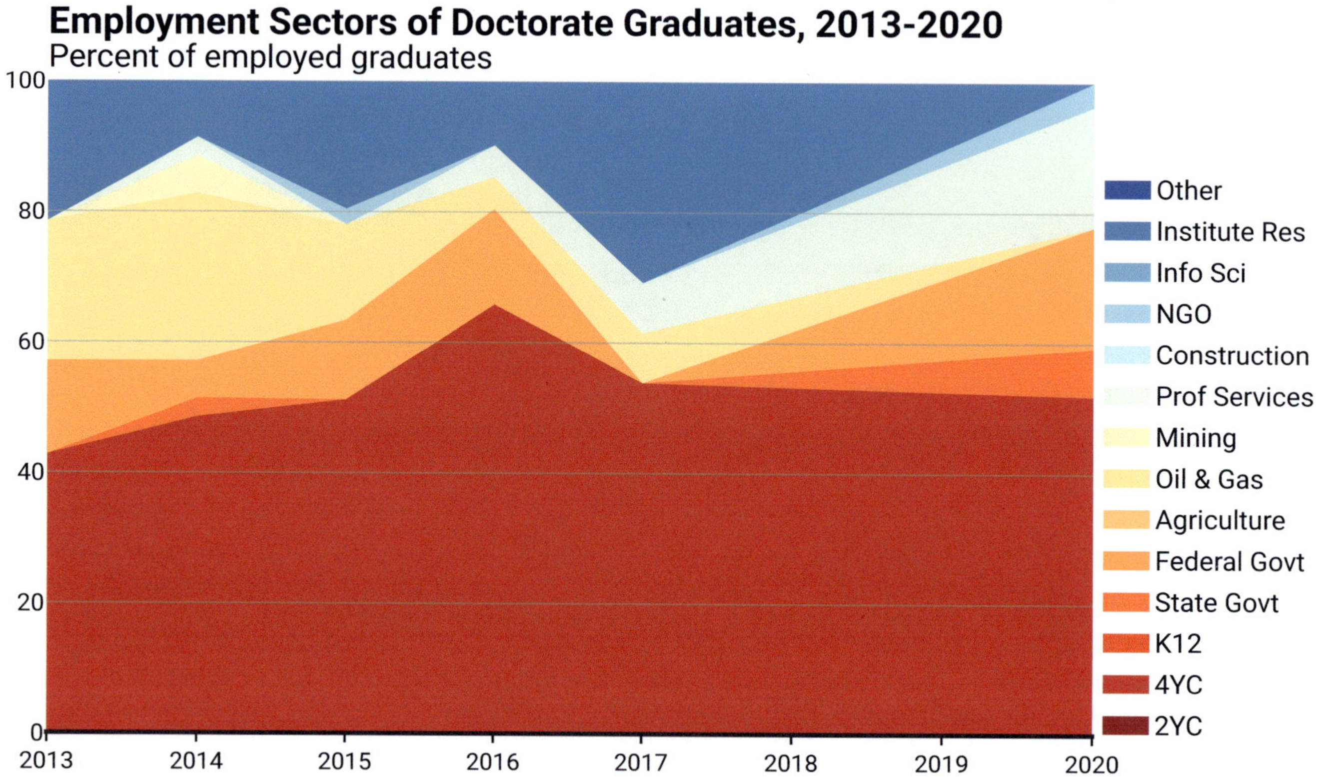

Universities and research institutes remain a steady source of employment for doctorates. These areas also include those graduates that are securing post-doc positions following graduation. More recently there has been an increase in hiring of doctoral graduates into federal government agencies to replace recent retirees, and interestingly, an increase in employment in the professional services sector—likely representing demand for doctorates' higher skill levels. Additionally, prior hiring in the oil and gas sector, which constituted almost a quarter of all doctorate hires about 10 years ago, has all but disappeared in most recent years.

4.16 **Multiple Job Offers**

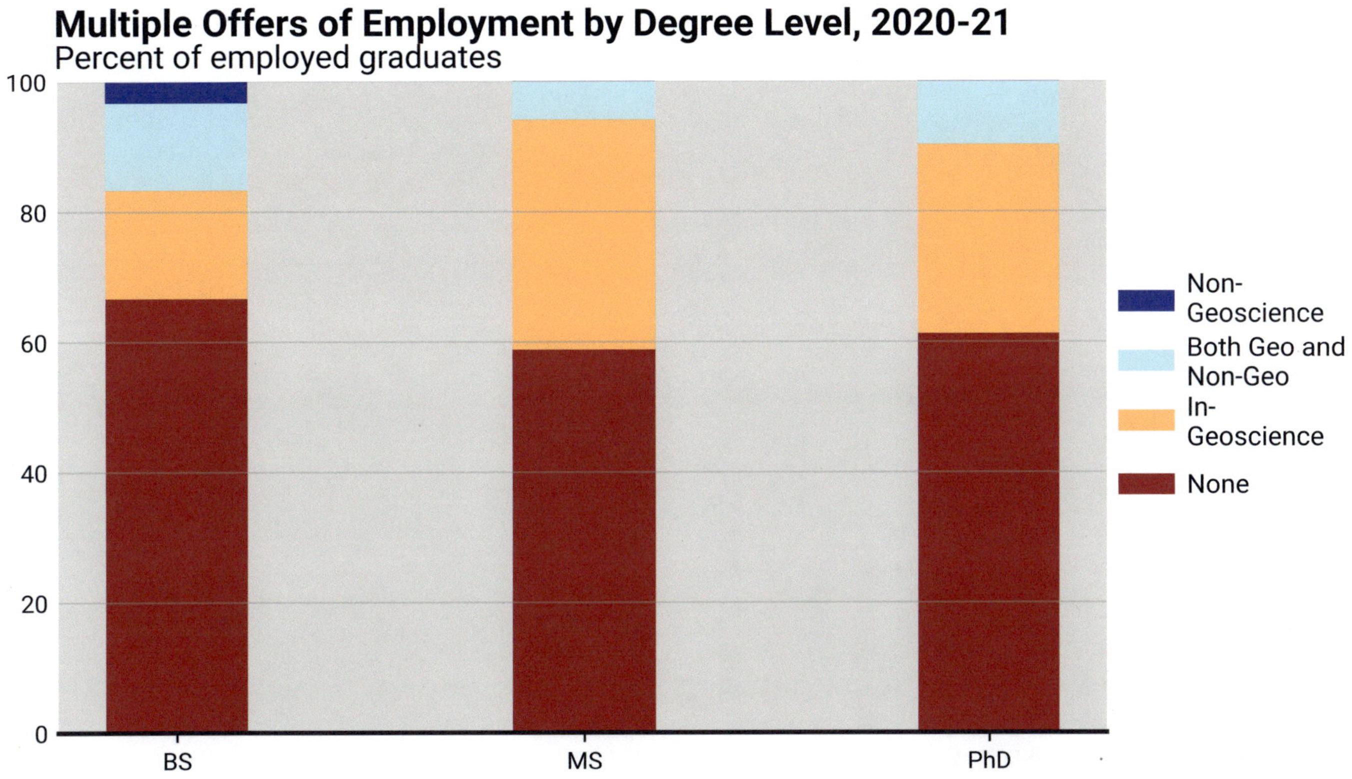

Of the graduates employed at graduation, approximately 40% had received multiple job offers, with a balanced split for bachelor's graduates between geoscience and non-geoscience occupations, while most graduate-degree recipients reported receiving multiple offers for geoscience occupations.

4.17 **Starting Salaries**

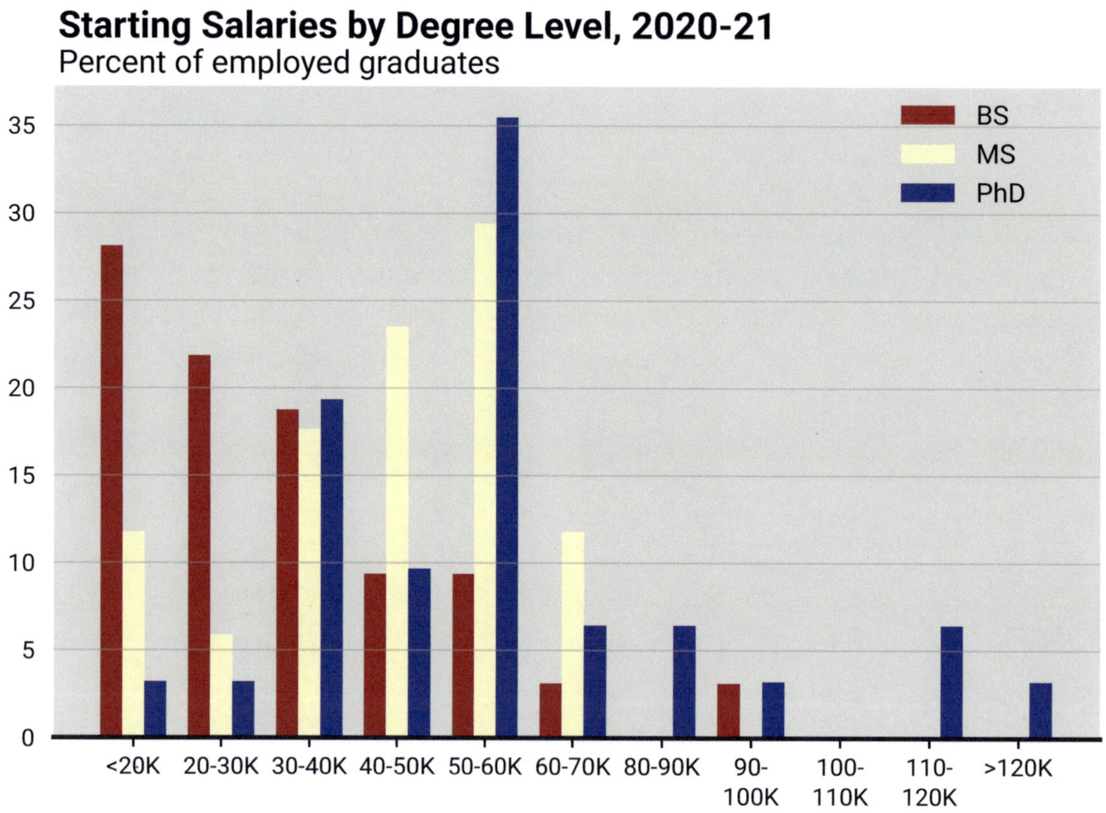

The starting salaries of bachelor's graduates have historically been lower than those with graduate degrees. This year, bachelor's graduates' salaries were particularly low, reflecting impacts from the pandemic with uneven employment quality and challenges in securing more traditional full-time employment. At the doctoral and master's level, there was a downward shift in starting salaries from prior years, which was mostly driven by weak hiring in the oil and gas industry.

4.18 Why Graduates Sought Employment Outside of the Geosciences

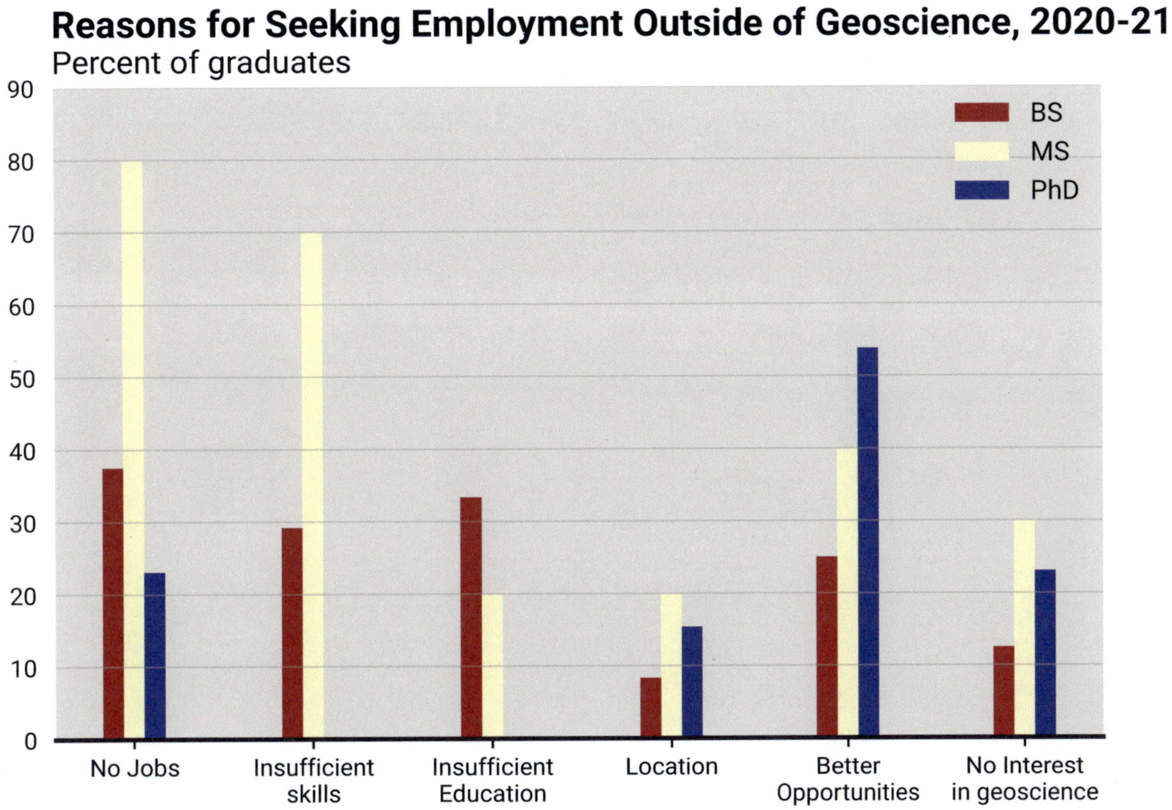

Most graduates sought employment within the geosciences; however, some decided to seek employment in non-geoscience opportunities. The two primary drivers to seek non-geoscience employment at the bachelor's and master's level were a lack of jobs and insufficient skills or education for the jobs that were available. Failure to identify geoscience jobs was potentially driven by an inability for graduates to relocate or they were seeking jobs in narrow occupational areas, such as oil and gas. For doctoral graduates and some master's graduates, the major driver was a recognition of better opportunities in other fields, reflecting skill transferability. About a quarter of the graduate-level respondents indicated a loss of interest in the geosciences.

Made in the USA
Monee, IL
05 November 2026

67378940R00048